# BEYOND
# HUMAN

**ALSO BY EVE HEROLD**

*Stem Cell Wars*

# BEYOND HUMAN

## HOW CUTTING-EDGE SCIENCE IS EXTENDING OUR LIVES

# EVE HEROLD

THOMAS DUNNE BOOKS

ST. MARTIN'S PRESS

NEW YORK

THOMAS DUNNE BOOKS.
An imprint of St. Martin's Press.

BEYOND HUMAN. Copyright © 2016 by Eve Herold. All rights reserved.
Printed in the United States of America. For information, address St. Martin's Press, 175 Fifth Avenue, New York, N.Y. 10010.

www.thomasdunnebooks.com
www.stmartins.com

Designed by Omar Chapa

The Library of Congress Cataloging-in-Publication Data
is available upon request.

ISBN 978-0-312-37521-8 (hardcover)
ISBN 978-1-4668-4294-6 (e-book)

Our books may be purchased in bulk for promotional, educational, or business use. Please contact your local bookseller or the Macmillan Corporate and Premium Sales Department at 1-800-221-7945, extension 5442, or by e-mail at MacmillanSpecialMarkets@macmillan.com.

First Edition: August 2016

10  9  8  7  6  5  4  3  2  1

# Contents

# BEYOND HUMAN

# 1

## When Humans and Technology Merge

Meet Victor, the future of humanity. He's 250 years old but looks and feels 30. Having suffered from heart disease in his 50s and 60s, he now has an artificial heart that gives him the strength and vigor to run marathons. His type 2 diabetes was cured a century ago by the implantation of an artificial pancreas. He lost an arm in an accident, but no one would know that he has an artificial one that obeys his every thought and is far stronger than the original. He wears a contact lens that streams information about his body and the environment to his eye and can access the Internet anytime he wants through voice commands. If it weren't for the computer chips that replaced the worn-out cells of his retina, he would have become blind countless years ago. Victor isn't just healthy and fit, he's much smarter than his forebears now that his brain has been enhanced through neural implants that expanded his memory, allow him to download knowledge, and even help him make decisions. While 250 might seem like a ripe old age, Victor has little worry about dying because billions of tiny nanorobots patrol his entire body, repairing cells damaged by disease or aging, fixing DNA mistakes before they can cause any harm, and destroying cancer cells wherever they emerge.

With all the advanced medical technologies Victor has been able to take advantage of, his life has not been a bed of roses. Many of his loved ones either didn't have access to or opted out of the life-extending technologies and have passed away. He has had several careers that successively became obsolete due to advancing technology and several marriages that ended in divorce after he and his partners drifted apart after forty years or so.

His first wife, Elaine, was the love of his life. When they met in college, both were part of a movement that rejected all "artificial" biomedical interventions and fought for the right of individuals to live, age, and die naturally. For several decades, they bonded over their mutual dedication to the cause of "natural" living, and tried to raise their two children to have the same values. Then, one day, Victor unexpectedly had a massive heart attack. Having a near-death experience shook him to the core, and for several years he and Elaine both pursued every natural avenue of fending off heart disease. They exercised, ate only heart-healthy foods, and Victor took a cholesterol-lowering drug. However, his heart disease gradually worsened, and by the time he was sixty-five, he had prematurely entered end-stage heart failure. Victor's heart had become grossly enlarged, and it was greatly weakened in the process. Day after day, he felt weak, dizzy, and had more and more trouble breathing. His feet and legs swelled up so much from water retention that he could barely walk. Then he could no longer sleep lying down because the fluid in his lungs made him feel like he was drowning. Being both ill and severely sleep-deprived made Victor's quality of life miserable. Elaine, who was in much better health, remained completely devoted to caring for him.

Gradually it dawned on Victor that he was dying. After all the years of illness and disability, this should have come as no surprise, but he was deeply disturbed by the idea. He and Elaine had a loving marriage and had just welcomed their first grand-child, and Victor's love and anticipation of seeing his grand-daughter grow up was far more intense than anything he had ever imagined. Soon another grandchild was on the way, and he wanted desperately to be alive long enough to welcome and know this child. He took stock of his situation. By then, there were millions of people who had received artificial hearts and thereby been completely cured of heart disease. Although he had always thought that he did not want to live to an advanced old age, he couldn't deny that he knew more and more people who had accepted some of the radically life-extending technologies becoming available, and had achieved far greater health and vitality than he and Elaine enjoyed. He had never accepted a pacemaker or an internal defibrillator, so his heart disease had proceeded unchecked and his health was rapidly deteriorating. Soon his cardiologist could do nothing more to assist him as long as he remained stubbornly attached to the worn-out heart he was born with. When he asked his cardiologist whether he would live to see his new grandchild born, the answer was, "Probably not."

His cardiologist disapproved of Victor's refusal to accept an artificial heart. Artificial hearts had completely replaced biological heart transplants because they could not be rejected by the body, were widely available, and were far more durable than biological hearts. So far, the earliest artificial heart transplants had already lasted more than eighty years, and the technology was constantly improving. Still, Victor was rather set in his ways and found the idea of having his natural heart removed and replaced by a metal and plastic electronic device deeply unsettling. Then one night, he woke Elaine up in a panic, telling her that he had severe chest pains and couldn't breathe. Elaine immediately called 911, but in the meantime Victor stopped breathing.

The next thing Victor remembered, he was in the hospital

emergency room with doctors, nurses, and emergency medical personnel swarming around him. They had revived him with repeated shocks from a heart defibrillator, but he felt his heart fluttering wildly and lost consciousness again. The next time he opened his eyes, his wife, son, and daughter were all gathered around him, their eyes red from crying, while his cardiologist was telling him something that he at first couldn't understand. He caught the words "terminal" and "surgery," being spoken with great urgency. Then he focused on the faces of his adult daughter and son as they leaned over him, their faces stricken and their eyes full of tears. The thought of never seeing these beloved faces again seemed utterly impossible to accept. With a weak, silent nod of his head, he agreed to the implantation of a permanent artificial heart. While Elaine signed the release form for Victor to have surgery, an anesthesiologist quickly administered an injection into his IV, and he drifted away again.

Victor's life after the surgery was remarkable. He suddenly had more energy and mental clarity than he had enjoyed in twenty years. In fact, it was only then that he realized how terribly sick he had been. The fluid in his lungs and the swelling in his body completely disappeared, and he told Elaine that he felt like an entirely new man. His long-held ideology about aging and dying "naturally" suddenly seemed stubborn and irrational. He noticed that even though Elaine was relieved and grateful that he was still alive, she wasn't changing her mind about her own dedication to allowing the aging process to proceed without any drastic intervention. He consoled himself with the secret belief that Elaine would change her mind when she faced her own serious health crisis. And he insisted that he still drew the line at getting one of the neural implants that so many people were hailing as a miracle cure for age-related memory problems, even Alzheimer's disease. It seemed that he and Elaine still had plenty of time ahead of them to enjoy their growing family, including four grandchildren who were rapidly approaching their teen years. It was hard to believe, but soon they would be entering adulthood, getting married, and having children of their own. Victor noticed that he had more energy and vitality than Elaine, who now had several chronic health problems, but he felt sure that all she needed was a "wake-up call" via some health crisis to convince her that it was time to take advantage of some of the amazing new medical technologies that would rejuvenate her and drastically extend her life.

There was a turning point for Elaine. She developed sharp pains in her lower abdomen and felt tired all the time. Victor urged her to go to the doctor, but she only became cranky and stubborn in her insistence that it was "only old age." She lost an alarming amount of weight and wanted to sleep seemingly all the time, so after a few months of Victor nagging her constantly, she finally went to see her gynecologist. There were a few tests that were swiftly followed by some devastating news. Elaine had stage 4 ovarian cancer, which had metastasized throughout her abdomen and had even entered her lungs and brain. Traditional approaches to treating cancer would do little to help her because the tumor in her brain was inoperable. However, her oncologist assured her that there was a good chance of curing the cancer by using specially engineered nano-sized particles that would seek out and destroy all the cancer cells in her body. Victor, who was with Elaine at this meeting with the oncologist, immediately latched on to the idea and heaved a huge sigh of relief. Then he couldn't believe the words that came out of Elaine's mouth. "I've lived long enough," she said. "I just want to go home to die. You can bring in hospice, but I don't want anything else done to me. Just let me die in peace."

Elaine's death was the hardest thing Victor had ever had to

face. She stuck by her decision to accept only palliative care, and within three months, she had passed away at home with their children and grandchildren around her. Her death was peaceful, but Victor was anything but at peace. His last days with Elaine were greatly complicated not only by grief but by an irreconcilable anger at her. He was unable to accept her decision to reject the nanotech cure that had already saved millions of lives. After almost sixty years of marriage, he felt that he couldn't go on without her by his side, and he fell into a deep depression. He then understood what it was like to want to die, and he even cursed the artificial heart that he felt had "sentenced" him to a long life without Elaine. He bitterly regretted that he had strayed from his original commitment to aging and dying "naturally." If only he had allowed nature to run its course, he never would have had what now seemed to him an intolerable stretch of years, perhaps decades, without his soul mate.

In the years after Elaine's death, Victor refused to entertain any possibility of marrying again, putting all his energy into their children and grandchildren. By then, one of his biggest problems was severe loss of vision due to macular degeneration, which was destroying the light-sensitive cells of his retina. He soon got to the point where he could no longer read, drive, or even watch a movie to ease his loneliness. He became more and more dependent on his daughter, and he felt guilty for the burden he believed he was becoming to her. He finally decided that he would embrace the microchip implants that would restore his vision, still telling himself that he was not artificially extending his life, only relieving his daughter of the burden of taking care of him. The microchips were miraculous. Not only did they restore Victor's vision to what it had been in his twenties, but being able to see and get around again gave him a new lease on life. He wanted to stop watching life pass him by and become active and engaged once again. He had been retired for twenty years, but now he felt that reentering the workforce would give him a focus, and he longed for a new opportunity. However, with his newly restored vision, when he looked into the mirror he saw an old man. He had even started to think of having a new partner in life, but what employer or woman would be interested in the wrinkled-up, elderly codger he saw in the mirror?

There was a new antiaging treatment that was being met with wild enthusiasm across the country. It sounded almost like science fiction. Doctors had devised an extremely "smart" treatment that released tiny nanoparticles into the body, where they entered every cell and "corrected" just about any problem, including the ubiquitous DNA mistakes involved in aging. People were claiming that the nanobot treatment literally erased all signs of aging. Victor had seen before and after pictures that were almost impossible to believe. He felt guilty when he thought about his shared commitment with Elaine to age and die naturally, but that option had already been nullified the day he accepted his artificial heart. If he was going to live several more decades, why not look and feel as young and vital on the outside as he felt on the inside?

A hundred years later, Victor finds himself once again ambivalent about the wide array of technologies that he has accepted to keep him young, productive, and fit. His closest companion is a robot that caters to his every need, yet leaves him nostalgic for Elaine and longing for a more authentic relationship. At times he feels guilty for having lived so long in a radically unequal world, where not everyone can enjoy life-extending enhancements, but should he be involved in a serious accident, his nonhuman parts are almost sure to keep him alive. If he wanted to die, no doctor would turn off the technology that is keeping him alive since doing so would be considered homicide. His only option is to discontinue his dependence on constant rejuvenation, then to age and die a complicated death as his bionic implants gradually go bad, a process that will take many decades and possibly, great suffering. At several junctures in his life, he regarded the technologies he has relied on as massively liberating, but as he continues to live decade after decade, they are beginning to feel like a trap.

While Victor's story may sound like science fiction, the technologies extending and enhancing his life are in fact now in development, and some are already being tested in humans. These technologies will radically transform human health and extend our life spans far beyond what most of us have ever dreamed. Many people alive today will be able to take advantage of an array of medical technologies taking shape at the nexus of computing, microelectronics, engineering, gene therapies, cognitive science, nanotechnology, cellular therapies, and robotics. The combination of these technologies is a nascent but rapidly advancing field that many scientists refer to as converging technologies (CTs). Scientists predict that combining the powerful emerging discoveries of today will take medical science, and human life, to an entirely new plateau.

Rather than predicting the effects of nanotechnology, genetic engineering, and cognitive science in isolation, experts say that one can only glimpse the true potential of these fields by looking at their combined effect. The results of collaboration among various scientific specialists is not only leading to an entirely new multidisciplinary approach to medical research, but creating treatments and cures that are far beyond what we now consider to be on the cutting edge. The possibilities for life extension will soon transform not only individual lives, but society as well. At the same time, the technologies discussed in this book will almost surely introduce ethical quandaries and complications that we are ill prepared to navigate. With a multitude of technological blessings come complicated practical and ethical issues, some of which we can predict, but many we cannot. Artificial organs and other critical body parts, neural implants to enhance the brain, nanorobots that can cure disease and reverse aging, and direct interfaces between our bodies and machines will dramatically improve human health, but they also mean that the line between "human" and "machine" will become progressively more blurry.

This book will tap the minds of doctors, scientists, and engineers engaged in developing these new technologies while telling the stories of the patients testing the newest products entering the market. It will ask difficult questions of the scientists and bioethicists who seek to ensure that as our bodies and brains become ever more artificial, we can hold on to our humanity. Will everyone have access to technological miracles, or will we end up living in a world of radical disparities? Will our descendants live in a world radically liberated by technology, or will we end up merely serving the machines and devices that keep us healthy, smart, young, and alive?

With radically advanced technologies come questions we have never before faced. Consider the simple development of a gene therapy that eliminates the tendency to become overweight. Today millions of people suffer a multitude of disorders that are directly related to obesity, including diabetes and heart disease. Yet many researchers say this is by no means inevitable. Soon drugs and supplements will prevent the excess absorption of calories. Dr. Ron Kahn at the Joslin Diabetes Center in Boston has gone even further. He has identified a gene that acts as a "fat insulin receptor" and turns sugars into fat, and he has learned to block the expression of this gene in mice. As a result, the mice are able to eat as much food as they want while remaining lean and healthy. Not only do they not get fat, they live 18 percent longer than the control mice. Such a discovery, if carried over into humans, will open up unprecedented questions. Should parents have this gene deactivated at a child's birth? Will we all choose to have the gene deactivated in our sex cells—eggs and sperm cells—so that the change will be automatically inherited by all of our offspring, changing the course of human evolution? Or will the gene deactivation be something that only each individual can decide for herself after having reached a certain age? The temptation to adopt this technology will be irresistible to just about everyone, yet there could be consequences that we haven't dreamed of, simply because such a possibility has never existed before. And not everyone is comfortable with transcending the "natural" limits that God or nature has conferred upon us.

According to the science writer Ronald Bailey, "Nothing could be more natural to human beings than striving to liberate ourselves from biological constraints."[1] But because we are human, that striving is poised to challenge our values and beliefs as never before. What is a reasonable life span for today's generations that doesn't adversely affect other generations? Is brain enhancement with drugs or neural implants "cheating" in the game of life? Who should receive advanced, expensive technology in the form of artificial organs and nanomedical interventions, and will everyone be equal in this new world of radical life extension? Or will the availability of advanced technologies create extreme disparities between those fortunate enough to live in modern, industrialized societies and those who don't?

Both left-leaning and right-leaning bioethicists have strong criticisms of proposed human enhancements, and some predict consequences as dire as internecine wars between the enhanced and the unenhanced. Past experience has shown that new technologies don't suddenly take over the world wholesale; technology adoption tends to come in waves, with the wealthy being the first beneficiaries while progressively cheaper forms of manufacturing make it available to all but the poorest poor. Yet this, too, is changing.

Advances in wireless computing, microelectronics, drugs, cell and gene therapies, nanotechnology, and robotics have reached a state in which all these fields are coming together in a new synthesis of treatments and technologies. The rate of change in these fields has become not just incremental, but exponential, as one technology builds upon another, extreme miniaturization occurs, and the cost of manufacturing plummets. As this process takes place, human life could extend for hundreds of years. Of course, in order to make such a long life meaningful, natural aging will have to be dramatically reduced so that those years are healthy, vital, and independent.

Already in use or in development is an array of artificial organs, including hearts, kidneys, pancreases, lungs, retinas, and parts of the brain. Those of us living today stand a good chance of someday being the beneficiaries of such advances. These human "components" are already saving and extending lives, but once nanotechnology, or the manufacture of tiny machines at the atomic level, becomes available, we will have entered an entirely new paradigm. At that point, microscopic nanoparts will be able to enter our bodies' cells and repair just about any damage brought about by aging, disease, and genetic mutations.

The last few years have seen wireless computing technology being integrated into a huge array of products, including our bodies, our homes, our gadgets, and our clothes. The broad distribution of computing will make our lives easier and more convenient than ever. Yet with this technology comes a profusion of intimate information about our bodies and brains that is likely to be stored on the Internet. Who "owns" that data, and who will have access to it? Can this data be protected from insurance companies, employers, and the like? And what happens when we no longer wish to be monitored? Will our doctors be willing to turn off artificial organs, for example, when we are ready to embrace the end of life, or will they consider such actions euthanasia?

There's also the possibility of a backlash against technologies that merge biology with technology, while many people continue to place a higher value on that which is "natural." The key issue is that many new technologies don't just cure disease—they go far beyond that and may end up enhancing almost every human ability. As these technologies advance beyond even current imagination, the changes to our lives, and the necessity to use them only for good, will require our best minds to literally rethink almost everything we currently know about being human.

Some people today have strong feelings against the quest for enhancement, but the pursuit of "perfection" goes back at least as far as the ancient Greeks, and arguably much further. Human beings have employed not only medicine, but cosmetic enhancement throughout history. Tattoos, piercings, cosmetics, and even some practices we would consider bodily mutilation have been common for centuries.

The search for the extreme enhancement of the human body and mind today fall under the rubric of transhumanism, a movement embracing the concept of the merging of humans and machines, that has been a marginalized system of thought in the scientific community for the past few decades. However, skepticism is giving way in the face of rapid breakthroughs in CTs, especially in light of rapid gains taking place in the private sector that are already in human trials. The strands of opposition from the right wing (especially religious conservatives) and the left wing (especially environmentalists and antiglobalists) have deep historical roots. Religious conservatives are especially concerned about what they see as hubris, a concern that also goes back at least as far as the ancient Greeks.

In ancient Greek mythology, the god Prometheus, who was thought to have created humanity from clay, made the error of sharing the ability to make fire with humans. As a result, humans became in a sense godlike, transcending their limits in ways that offended the gods. As a punishment for his crime, Zeus had Prometheus bound to a rock where every day an eagle would come and feed on his liver. Overnight his liver would regenerate itself, only to be eaten again the next day. Prometheus was condemned to an eternal, tormenting punishment for trying to raise mankind above its natural state. The message of this myth reflects a deep dread of catastrophic punishment should we attempt to transcend the "natural" limits of humanity. The theme that man would suffer some dreadful punishment if he attempts to better his lot has been reiterated endlessly throughout Western history and has numerous supporters even today. Bioconservatives today often invoke the charge of "playing God" in a whole array of medical and biotechnology issues and remain convinced that should humans, in their hubris, cross some line that enables them to develop "godlike" powers, the unintended results would be catastrophic.

Transhumanism has been so deeply intertwined with human history that the seeds of transhumanist thought can be seen since the beginning of time. The transhumanist trend also has deep roots in rational humanism and the Enlightenment, and its sweep seems to grow wider with each passing year, thanks to scientific innovation. The deep longing for perfection that gave rise to transhumanism can be seen in everything from Plato's *Dialogues* to medieval Christianity to modern science fiction. However, the vast majority of doctors, researchers, and others in the field of developing and regulating CTs adamantly reject the label "transhumanist" because of its radical connotations.

The term "transhuman" was first coined in 1927 by Julian Huxley, a biologist and brother of author Aldous Huxley. Huxley used it to denote the human organism in transition from "human" to the next stage of existence, described as posthuman. In this book, I use it simply to signify humanity in an active process of profound change, and not necessarily in the strictest sense described by the World Transhumanist Association (WTA). The WTA represents the contemporary Transhumanist social movement, or Transhumanism with a capital *T*. This movement began to form in the 1960s and, although it might appear to have languished in the obscurity of a small community of unconventional thinkers, it's gathering strength from the amazing trajectory of scientific discovery. This book is about transhumanism with a small *t*, i.e., concrete developments that are taking place today, embed human beings more deeply with technology than ever before, and call for a deeper examination of these changes.

The international nature of convergent technologies and the issues surrounding them is important because so many of the innovations are being developed not by governments but by private industry in nations that span the globe. New and convergent technologies are being commercialized and spread by companies that are themselves becoming global. However, regulatory and ethical policies differ from one country to the next. What is considered acceptable in China may not be acceptable in the United States or Britain, and vice versa. So far there is no all-encompassing system of regulation that recognizes a shared system of values, or that ensures that people throughout the world will have equal access to the best of technology.

Francis Fukuyama has referred to transhumanism as "the world's most dangerous idea." In a famous 2004 essay, he asserts that "society is unlikely to fall suddenly under the spell of the transhumanist worldview. But it is very possible that we will nibble at biotechnology's tempting offerings without realizing that they come at a frightful moral cost. The first victim of transhumanism might be equality."[2] Like our fictional character Victor, we are likely to gradually accept more and more technology at junctions in our lives when we are particularly vulnerable to making choices that will avert some crisis. But Fukuyama writes as though it is a foregone conclusion that unenhanced, unhealthy persons in the future will be unequal at their very essence. If some portion of society grows stronger, smarter, healthier, and longer-lived, he assumes that the concept that all human beings are equal will be swept away, along with the rights of the unenhanced. However, it is not at all clear that society will not delineate an even deeper quality that transcends health and ability that forms a newly recognized nucleus that confers human dignity. Or, conversely, that the same core attributes (yet to be fully defined) that make us human are unchangeable no matter how many technological enhancements we accept.

Central to this debate is the question of what composes human identity. Does our identity change when parts of our bodies become artificial? What about if we augment our brains with expanded memory or computational powers and learn to communicate wordlessly to other people through direct brain-to-machine interfaces? While philosophers have argued throughout the ages on how to define human identity, CTs add an extra sense of urgency to the issue. Is human identity a fixed value that cannot change, or is it an evolving phenomenon for each of us as individuals as well as for the whole human race? Ideally, we would have settled this question before we tread farther into the territory of enhancing our bodies and brains. However, as has always been the case, technology leapfrogs over philosophical questions at an ever-increasing speed. It's entirely likely that it is technological enhancement that will over time reveal to us who we are at our center, rather than being some quality that is pre-determined by us in advance. Ironically, we may be able to define ourselves only after we have artificially enhanced our bodies and brains to a higher level of functioning.

Proceeding alongside the development of technologies to en-hance human abilities are amazing breakthroughs in the devel-opment of robots and artificial brains. It's likely that our destiny will be shared with increasingly intelligent and capable machines that may even surpass our own intelligence. We are destined to interact extensively with these robots as even today they are be-coming more and more humanlike in their ability to do every-thing from rendering basic health care for the sick and elderly to learning how to identify and respond appropriately to human emotions. Caregiving robots for the elderly are just around the corner, and these robots are being designed to "learn" from their interactions with their owners so that they will develop unique and intricate relationships with them. Not only will we depend on robots for myriad functions, some of them emotional, ad-vanced robots will take over more and more human jobs.

While, as noted above, technological advancements are tak-ing place at a faster pace than the moral, legal, and regulatory control of them, this very fact is placing immense pressures on scientists, philosophers, and legal and medical personnel who are working at the leading edge of CTs to delineate their ethical uses. Massive efforts to analyze and direct such technologies are be-ing made by the U.S. government, the European Union, and many other countries. In addition, the as-yet-immature field of bioethics, which barely existed a generation ago, is gaining a sure foothold as a part of medicine and in the development of new technologies, and it now has a place at the table when governments convene to consider public policies. In fact, bioethics is a rapidly growing field that cannot be dominated by machines. The real task before us is to move these debates into the mainstream so that more people from all segments of society are participating in the conversation. All of these factors represent steps in the right direction, but it is still probable that even the smartest people on the planet can't predict some of the side effects and consequences that could result once powerful, life-changing technologies are unleashed.

Although it is widely acknowledged that bioethical explorations are integral to the advancement of medical technologies, there is another assumption at the heart of the antitechnology argument that calls for examination. That assumption is that human nature, whether created by God or evolved by nature, is a fixed, immutable phenomenon incapable of learning from the mistakes of the past. Antitechnology thinkers tend to assume that (1) humanity is a finished product (something that evolutionary biologists would take issue with) and that (2) the level at which we make decisions today is the same level we will be at in the future—i.e., that as we expand human knowledge and ability, we will not reach new plateaus of understanding that will enable a more advanced process of decision making. In fact, we can't know today what decisions our progeny will make hundreds of years from now, or assume that those decisions will be as inept and problematic as many decisions made in the distant past. We must also consider the evidence that human beings are capable of learning from past mistakes and that cultures and societies may be advancing toward more enlightened, democratic systems that tolerate and include more human diversity than ever before.

One phenomenon that will have a profound influence on future decision making will be the presence of millions of people in society who are physically young and vital but who have the wisdom of having lived many more years than is currently possible. Throughout human history, societies have been largely impoverished of the kind of wisdom that only life experience, gained over many years, can provide because most people had very short lives. There have always been a few individuals who lived until old age, but the mass of humanity has not. Just consider the fact that in the United States in 1900, the average life expectancy was 47, and today it's 78.8 years. In addition to being few and far between, very old people in the past were at the mercy of manifold disabilities due to age-related diseases like heart failure or Alzheimer's disease. One very important open question about radical life extension is whether we will be able to cure or prevent the common age-related conditions that even today reduce the quality of life and introduce devastating disabilities in those over age sixty. There is a widespread consensus that just prolonging life in an advanced stage of deterioration is not a desirable goal.

Bill Joy, the cofounder of Sun Microsystems, has pointed to some of the dangers inherent in some emerging technologies, such as robotics, nanorobots, and engineered organisms that will have the ability to replicate themselves. Once self-replicating machines are set into motion, will they replicate to the point where they overwhelm the human population by utilizing all of the natural resources? Self-replication of machines will have to be controlled just as human reproduction will have to be, or we will soon overwhelm the planet's ability to support huge numbers of long-lived human beings. Another danger Joy presents is that with the ubiquitous nature of knowledge on the Internet, the information to make potentially immensely powerful technologies will be available to everyone, including malevolent individuals who would like nothing more than to wreak havoc.

Joy estimates that self-replicating robots will be possible by 2030. Imagine that a robot programmed for malevolent purposes creates a large number of copies, all networked together and focused on the goal of eradicating humanity. In this case, after having subdued dangerous animal predators, humans will be faced with a struggle with machine predators that may be extremely powerful. Joy is especially concerned about self-replicating nanorobots, the molecule-sized machines that could vastly improve human health and fight aging, escaping into the environment and sucking up all of the earth's resources in the process of replicating out of control. Scientists and bioethicists refer to this as the "gray goo" problem, a scenario in which the world's resources are subsumed into a mass of self-replicating nanobots. Considering that nanotechnology is rapidly advancing, this is an urgent matter to consider. However, there are nanotechnological solutions available even today, such as designing nanobots that can be turned off once they have completed their task—such as destroying cancer cells—or made of the same type of polymers that dissolving sutures are made of. These nanobots have a time-limited existence that precludes the ability to build up in the body or the environment. It is critical that scientists universally protect against the gray goo problem as they design all nanobots.

One cause for concern that Joy highlights should truly give us pause. That is the fact that CTs are mostly being developed by corporate entities whose prime purpose is the rapid commercialization of new technologies. While it is standard these days that such companies have an oversight committee that includes bioethicists, private entities can never be completely controlled by government oversight bodies such as the U.S. Food and Drug Administration (FDA). Private companies have a strong incentive to be the first to patent and commercialize new technologies, and this means they have a disincentive to share data. It's worth asking whether the for-profit business model is the most appropriate one in the case of such radically transforming technologies. The question of sane and reasonable regulation should be paramount, and in a global market, universally shared. To ensure the universal adherence to safe policies and procedures, the economic field must be leveled by common international rules and regulations.

Many people alive today will have to personally grapple with the side effects of CTs, both the good and the bad. For example, scientists very recently achieved a direct interface between a human subject's brain and a computer by capturing the electrical pulses of the subject's brain and converting them into signals that the computer could read. The human subject sent a very simple message, "move a certain finger," across the Internet to another computer that was connected to another person. The recipient got the message and moved his finger as instructed. As rudimentary as this exercise was, it proved that human brains can be connected to both computers and, through them, to other brains. While this procedure has tremendous possibilities for people to communicate directly with each other despite vast distances, the technology could be gravely abused. With direct brain-to-network and brain-to-brain connections, could an employer, for instance, require each employee to connect to the "company network," a shared consciousness that not only intrudes the company's messages into the mind of the "user," but also potentially reads the user's thoughts? What essential rights to privacy do we all share regarding the nature of our thoughts? Will the right to keep our thoughts to ourselves be abdicated once this technology develops to a more sophisticated stage? Will governments and powerful corporations have the ability to "plant" thoughts in our heads?

For some people, there is something inherently unsettling about being totally dependent on artificial technologies for life or for a vital function. It's not at all clear whether artificial parts such as artificial organs or neural implants will be considered part of the recipient's body or a foreign object placed into it. Given the pace of development of artificial implants, the issue is greatly complicated by the increasing integration of implants with the body's natural processes. That being the case, we are going to be met time and again with the question of what is "you" and what is "not you."

This is especially critical at the end of life. For example, a technology that stimulates the endless production of stem cells from within the body could cure a wide range of diseases—or at least hold them at bay for a very long time. Yet the day will come when some incurable disease process reduces the quality of life to something less than desirable, perhaps even unendurable, and then we would want to put a stop to the generation of stem cells. When and how would the process be stopped, and who would stop it? Would cessation of this therapy be regarded as murder, suicide, euthanasia, or something else that we have yet to name? As much as we don't want to consider an end to human life, so far no one has escaped death. Some thinkers, such as Ray Kurzweil, predict that death will be eradicated when we succeed in uploading our minds into machines, but so far we are a long way from achieving immortality. Most of us, when given the choice, would gratefully accept many more decades of life if they were fit and healthy years. But the hope for biological immortality may ultimately prove futile. The fact remains that there is a compelling need to lay down some rules for the discontinuing of advanced technologies, including those implanted into our bodies, in order to allow a compassionate death.

Some questions may be unanswerable at our current state of technology. If we were to embrace human enhancement in every way, shape, and form that scientists and engineers can devise, would we still come up against a natural limit in life span, memory, or strength? Would we simply meet with the law of diminishing returns—continuously patching up a body that is so worn out and diseased that life becomes an unbearable burden? Do each and every one of us have a right to die at some point, even if it means deactivating some technology that is deeply integrated with our bodies? At the same time, might there be a nonbiological stage of existence that we would accept over death, and if so, what would it look like?

An intriguing question is whether there is some part of us that can't be enhanced through technology and can only evolve on the basis of lived experience. I am thinking here of mental, emotional, and spiritual phenomena, which may be forever beyond the reach of technology to alter. If this is the case, all technologically advanced enhancements applied to an immature being might bring very little change in one's quality of life or depth of experience. It's possible that we may never be able to answer this question until we embrace human enhancement on every level and see where it leads us.

The transhumanist philosopher Anders Sandberg has written that "we change as humans not because we are unhappy about who we are, but rather because we want to become better. Self-transformation is not a search for some imaginary state of perfection, as is sometimes suggested, but rather an open-ended process. As we grow as people our ideals and values also grow and change."[3] It's because of the open-ended nature of our perennial search for betterment that predictions of what human beings will be like one thousand, five hundred, or even two hundred years from now is so difficult. Those values that we cherish today may be of little importance to human beings who live on an experiential plateau far above the one we share today. On the other hand, those values that persist through the coming technological age will achieve greater validation. For example, if everyone becomes tall, thin, and beautiful through genetic engineering, being beautiful may no longer be the sought-after value it is today because it will be commonplace. It's possible that our values might shift toward a different standard of beauty or to something altogether different. The value that we currently place on inner beauty—a far more elusive concept—may survive and even intensify. The question is, will converging technologies fundamentally transform human beings and human life in a qualitative way, or simply entail a multiplying of qualities we already possess?

Bioethicists who write about the technological advances discussed in this book almost to a person call for the need to bring certain enduring (if evolving) values into sharp focus before we accept changes that could threaten those values. Each of the techniques presented here has at least the potential to threaten some of our most cherished values, including human dignity, freedom, equality, individuality, democracy, privacy, and autonomy. Some of them can threaten our abilities to make our own decisions about issues as intimate as the manner in which we die. Questions arise about the extent to which we have the right to embrace, reject, or limit profound changes to our bodies and brains. In a world where greatly augmented abilities in addition to a very long life are embraced by a majority of people, will one really have the option of choosing not to accept enhancement? Suppose that a significant minority of people either can't afford or decides to reject technological enhancement. Will those people represent a disadvantaged underclass? How will they be treated by the rest of society?

As illustrated by Victor's story, the questions we face not only entail a greater interaction with our machines (such as the way we interact with our cell phones), they entail an actual merging with our machines to the extent that our very identity is open to interpretation. A group of bioethicists led by Courtney S. Campbell, writing in the *Cambridge Quarterly of Healthcare Ethics* in 2007, say that, "At some point, perhaps, the 'other' that is machine will be increasingly difficult to differentiate from the 'self' that is embodied. Rodney A. Brooks, director of the Artificial Intelligence Laboratory at the Massachusetts Institute of Technology observes: 'While we have come to *rely* on our machines over the last 50 years, we are about to *become* our machines during the first part of this millennium.' "[4] The implanting of computer chips and electronic arrays that are poised today to restore vision, hearing, movement, and memory to those who are impaired will no doubt evolve into technologies that expand the senses and extend memory and learning far beyond our present "normal" abilities. One reason this is considered inevitable is because "normal" is a concept that eludes definition.[5]

While scientists and philosophers continue to debate the definition of "normal," it is quite possible that our notion of normality will expand to accommodate the widespread acceptance of performance-enhancing technologies. If biomedical advances proceed in the future in the way they have become integrated and accepted in the past few decades, a majority of people are likely to eagerly embrace technologies like retinal implants, functional electrical stimulation of paralyzed muscles, artificial organs—including artificial brain structures such as the hippocampus—cochlear implants, and a host of new abilities achieved through direct brain-to-computer interfaces. The reason this seems likely is that these technologies are now being introduced as lifesaving, life-extending repair mechanisms. On the basis of, say, restoring sight to a blind person or memory to an Alzheimer's patient, our sense of mortality and dread of death and disability is perhaps the most compelling drive of all. If curing blindness is a good thing, then extending human vision to encompass infrared light must be an even better thing. One need only to consider how multiple advanced technologies developed in the process of military research (blood transfusions, for example, were developed on the battlefield) have historically passed easily into civilian use. In a situation of war, would we hesitate to allow our soldiers such advantages as goggle-free infrared vision, supersharp hearing, and increased memory and concentration? And once these technologies become widespread in the military, what's to stop them from becoming widespread among the civilian population?

That which we accept in a health care setting is almost certain to bleed over into the territory of enhancement. What parent, if given the choice, would deny their children a gene therapy that would silence certain aging genes and extend their life and health dramatically? The line between "normal" and "better than normal" is already blurry; as biotechnology incorporates more artificial interventions into the very cells and tissues of our bodies, where will the boundary between "natural" and "artificial" lie?

As we grow to incorporate more artificial interventions in our brains and bodies, we will greatly expand the extent to which we interface with an environment that is "alive" and "intelligent," having incorporated wireless computing, GPS, and other technologies into virtually every common object of consumption. We will communicate effortlessly with our "smart homes," which respond to our presence through biological sensors and computer chips implanted into our bodies, self-driving cars, utilities and environmental safety features that respond automatically to our presence, and wearable computing that continuously monitors our vital signs, needs, and even moods. These technologies, along with bio-logically based identification techniques such as retinal scans and implanted identity chips, will make day-to-day life much simpler, just as the environment becomes much "friendlier." It's likely that we will quickly become so dependent on a smart environment that we'll wonder how in the past we lived without it, just as today we marvel that we ever lived without e-mail and smartphones.

But there's a risk to the computerization of everything, in-cluding our bodies and brains, and that is an unprecedented dan-ger to our privacy and autonomy. That is because all this information about us, our actions, our habits, and potentially even our thoughts will be digitized, recorded, and stored some-where. If our "smart" garments are able to continuously inform our doctor about our vital signs, that information must be stored and could potentially be hacked into. In a market-based econ-omy, it might even be sold to insurance companies and other businesses, or to a potential employer. And it may be made freely available to the government if we don't substantially regulate and place limits on who can have access to our personal infor-mation. Soon a huge, minutely detailed body of information about us will exist, and there's no guarantee at present that we will have control over that information. This is one compelling reason why some people may choose to opt out of many of the new technologies, to the extent that is possible. Still, accepting mul-tiple monitoring technologies will be the path of least resistance and in some cases may even be unavoidable. Suppose, for example, airport security systems come to rely on a required universal, implanted "identity chip" that must be read and cleared before a passenger can board an airplane. Embracing some technologies may be utterly mandatory if one is to live a normal life.

Through converging technologies, we're on a path to elimi-nating a huge amount of human suffering, but not everyone welcomes this idea. One of the major concerns of bioconservatives is the presumed nearly complete eradication of suffering on both a physical and a mental level. Throughout history, much has been written about the potential of suffering and adversity to build laudable character attributes, including compassion and sympathy for others. The British bioethicist Y. J. Erden notes that even the experience of depression can have a transcendent value when processed by a person who is self-aware and truth seeking. She writes that "our identity is formed by a multitude of experiences or emotions that may cause us pain, such as depression and bereavement."[6] Catholic ethicists in particular tend to believe that there is redemptive value in suffering, and these thinkers stand in stark contrast to what Erden calls an excessive "technological optimism" on the part of scientists. There is also a noticeable tendency on the part of the scientific community to think of the human experience in reductionist terms, as the sum total of chemical and electrical processes of the physical brain, and an attendant expectation that most human suffering can be ameliorated, and furthermore, ought to be eliminated.

It's hard to differ with such an argument, but it is also unproven that the elimination of diseases and the augmentation of human abilities will necessarily provide the elusive state of happiness that we humans perpetually seek. Science has proven that we invariably assimilate each improvement into our lives, which then becomes a "new normal," pushing the boundaries of what constitutes happiness ever outward. It seems unrealistic to assume that even in a world where disease was eradicated and everyone lived forever, life would be free of all adversity, including disappointment, frustration, loneliness, and the craving for meaning.

Most of the groundbreaking technologies described in this book are either in human clinical trials, in animal trials where they have already provided proof of principle, or already exist and simply await commercialization. Governmental regulators and academic and private-sector ethical oversight committees are struggling to catch up with the technology and its implications for individuals and society. Because these technologies have tremendous commercial potential, the speed of implementation is bound to occur very quickly in the United States, Europe, and several Asian countries, all competing to secure the first patents and dominate international markets. Regulatory bodies such as the FDA are moving at a snail's pace compared to the rapid development of medical technologies. Humanitarian concerns are putting even more pressure on governments to help make such technologies available to their sick citizens. They have barely begun to consider the issue of human enhancements.

The philosopher Nick Bostrom calls attention to "the distinction between enhancements that offer only positional advantages (e.g. an increase in height), which are only advantages insofar as others lack them, and enhancements that provide either intrinsic benefits or net positive externalities (such as a better immune system or improvement of cognitive functioning). We ought to promote enhancements of the second kind, but not enhancements that are merely positional."[7] But in a society that currently obsesses about youth and beauty, it seems unlikely that we will limit ourselves to only the second kind of enhancements.

While radical new inventions such as artificial organs are now being tested in humans, artificial intelligence (AI) in the form of robots that interact with human beings is relatively advanced and is about to become ubiquitous throughout the industrialized world. In the foreseeable future, we may be sharing our planet with huge numbers of intelligent, self-acting artificial beings whose intelligence (in terms of computational power) exceeds our own. In democratic societies, what rights and responsibilities will these beings have, and in what ways will they both resemble and be unlike us?

We're rapidly approaching the point where very few (if any) individuals understand the sea of technologies that we find ourselves immersed in. When supersmart machines take over most of human labor and dramatically extended life spans mean that people have a surplus of time on their hands, will we become lazy and dependent on the machines we no longer understand? How will we ensure that highly developed AI will serve the needs of humanity rather than its own? Some extremists think that we should abandon the development of AI, nanotechnology, and genetics because we will eventually create a future in which human beings are redundant and unnecessary. The neo-Luddites may be able to somewhat slow the adoption of innovations, but they can never win in a global society where technology is driven by competition in both the business and military sectors. Those nations that decide to sit out the coming revolution in CTs will be economically disadvantaged and militarily vulnerable.

Bostrom envisions a world in which transhumanism is rapidly becoming a reality. He writes, "Virtual reality; preimplantation genetic diagnosis; pharmaceuticals that improve memory, concentration, wakefulness and mood; performance-enhancing drugs; cosmetic surgery; sex change operations; prosthetics; antiaging medicine; closer human-computer interfaces: these technologies are already here or can be expected within the next few decades. The combination of these technological capabilities, as they mature, could profoundly transform the human condition."[8] This was written in 2005, and since then, each of these technologies has become a reality and merely awaits commercialization in order to become the mundane stuff of everyday life.

Anders Sandberg notes that the ability to change and augment our bodies, far from creating armies of identically "perfect" humanoids (as is frequently depicted in science fiction) is more likely to be exercised in intensely personal ways. New technologies, and the freedom to use them as each individual chooses, will take self-actualization to unprecedented levels. This movement, rather than imposing oppressive pressures to conform, could call upon society to tolerate more human diversity than we can currently imagine. It all depends not only on us, but on who we become as humanity 1.0 rapidly climbs the technological ladder that will bring about a new race—the transhuman race.

# 2

## "Better Than the Heart I Was Born With"

When forty-year-old Stacie Sumandig was told by doctors that she had only a few days to live, she had one thought going through her head. So sick that she didn't have the strength to cry, she thought, "I have four kids to take care of. Death is not an option." The news that a virus had attacked and destroyed both ventricles of her heart, leading to end-stage heart failure, was a shock that she simply couldn't process. Even though she had been feeling unwell for months, she was fit and active. In fact, up to two weeks before landing in the hospital, she had been working out three times a week and working full time at a landscaping job. Between a physically demanding job and caring for her young family, Stacie "almost never sat still."

Stacie's story proves that, no matter what doctors say, there is no substitute for listening to your body. She first went to her family doctor in July complaining of flu-like symptoms, night sweats, and difficulty breathing. She had a history of frequent bouts with bronchitis, and that is what her doctor diagnosed. She was prescribed antibiotics, which seemed to help for a while, but then the symptoms returned. Between July and September, Stacie had several visits to her doctor, first being diagnosed with bronchitis, then with pneumonia. She was told that her trouble breathing was probably due to asthma. While course after course of antibiotics seemed to temporarily help, the symptoms inevitably came back. Finally, in September, her doctor ordered a chest X-ray. Reading the results early one morning, her doctor picked up the phone and gave her a call she will never forget.

"Your heart is severely enlarged," the doctor told her, "and it's surrounded by fluid. You need to go to the hospital immediately." So began a journey for Stacie that would call for the courage and endurance that few people ever have to draw upon. Doctors at the hospital told her she would likely not live through the weekend. They advised her to get her affairs in order, to notify loved ones, and prepare for the worst. Rather than accepting the news, she could think only of her children, ages eight, nine, thirteen, and fourteen. Who would take care of them? How would her husband manage without her? "I was so sick that the tears wouldn't come out, but I told them, 'There has to be something you can do.' "

Soon it was decided that, as a last resort, Stacie should be transferred to the University of Washington Medical Center (UWMC) in Seattle, where there were more resources than at her local hospital in the town of Puyallup, Washington. Luckily, there she met a cardiologist named Dr. Nahush Mokadam, and he offered her last, and only, option. First, he told her that there was no biological heart available for a transplantation that could save her life. However, there was a very new alternative—the Total Artificial Heart (TAH), made by a Tucson-based company called SynCardia. Only a few hundred people had ever been implanted with the artificial heart, but the FDA had recently approved the technology and UWMC had the capacity to offer it to Stacie.

The TAH was not considered a permanent solution, but a bridge technology that would keep Stacie alive until a suitable biological heart became available. Even then, Stacie had a complication that meant that after her natural heart was removed—an irreversible act—she had only a fifty-fifty chance that the TAH could be attached to her aorta, the main trunk of the heart's arteries that carry blood away from the heart and out to the rest of the body. Stacie had been born with a congenital narrowing of her aorta. Her first heart surgery, a reconstruction of the aorta, was performed when she was only four years old and, at thirty-one, she had had an aortic valve replacement. Due to the damage from the virus, which doctors estimated had been attacking her heart for two years, the only thing holding her aorta in place was an infected, fragile piece of muscle that was attached to the aortic valve. It was entirely possible that once this small piece of flesh was removed, there would be nothing to attach the artificial heart to. Knowing that once she was put to sleep, she had only a 50 percent chance of ever waking up, Stacie was terrified, but the chance that the surgery would be successful was all she had to cling to. It was an enormous amount of information to process within a few days' time. The idea of having a totally artificial heart was more than a little unsettling, but thinking of her children, she agreed to go through with the operation.

It turned out that one of the effects of Stacie's long-term heart failure actually improved her chances of a successful implant. According to Dr. Mokadam, when he removed her heart, rather than consisting of the firm, elastic muscle that characterizes a healthy heart, her heart was so soft and mushy that it just slipped through his hands. It was also dramatically enlarged, and as a result, it had enlarged her chest cavity as well. This was one of the only reasons why Stacie, as an average-sized woman, could accommodate the SynCardia 70-cubic-centimeter (cc) heart, which is sized only for men or large-sized women. Fortunately, the TAH, which includes right and left atria, or upper heart chambers, and right and left ventricles, or lower chambers, which carry most of the work of the heart, could be attached to Stacie's aorta. But by then, Stacie's vital organs had begun to fail due to the lack of blood flow, and her lungs collapsed during surgery. She was placed in a two-week, medically induced coma, and connected to an ECMO (extracorporeal membrane oxygenation) machine that took over the circulation of her blood while her vital organs came back online and her lungs started to function again.

After two and a half weeks, Stacie woke up from the coma. Her vital organs were steadily improving due to the increased blood flow, and to her relief, the TAH proved to be "better than the heart I was born with." Even though she had two tubes coming out of her abdomen attached to the Freedom portable driver, a 13.5-pound apparatus, which she carried around with her in a backpack, she was able to go home and live a relatively normal life while she waited for a matching biological heart. The backpack seemed heavy at first, but she soon grew accustomed to it and the tubes attached to it, which were necessary as long as she had the TAH. The battery in the driver, which kept the heart pumping, had to be recharged every day by being plugged into a wall socket. The battery charge lasted for only eight hours, so she had to be ever mindful that an electrical outlet was always within reach. This in itself caused a certain nagging anxiety, but fortunately, she could keep the driver plugged in at night while she slept.

The upside was that the heart pumped 9.5 liters of blood per minute. Her kidneys started to work again and she suddenly had energy. The color came back into her face. Life flowed through her. She walked the dog, took care of her kids, and went shopping at the mall. It felt like a miracle.

Not everyone responded to Stacie's condition with sensitivity. People tended to stare at her, with the heavy backpack and the tubes that led from her backpack into her abdomen, and some of them asked "rude, obnoxious questions." The heart had the audible sound of an amplified heartbeat. Stacie quickly got used to it, but not everyone tolerated it. When she was well enough, she and her family went to the church they had attended before her surgery, a church with a sign outside that said "Come as you are." After attending one service, the pastor told her that the sound of her heart was bothering some of the other parishioners, and he told her not to come back. Stacie and her husband could hardly believe it, but they agreed not to return.

After 196 days of living with the artificial heart, Stacie got the phone call telling her that a donor heart had become available. She had intensely mixed feelings about the news. On the one hand, she was incredibly excited, but on the other hand, she was worried about removing the artificial heart that worked so well that she had been living a virtually normal life. What if her body rejected the biological heart? With the TAH, rejection was not an issue. She had to take medications to avoid blood clots, but having a donor heart would mean having to take lifelong drugs to suppress her immune system, a process that carries serious risks. The immunosuppressant drugs meant that a simple infection could potentially take her life. She went through with the biological heart transplant and today Stacie is an active, healthy, full-time mom. She has to have regular biopsies to monitor for rejection of her new heart, and she was recently surprised to learn that she was actually experiencing mild rejection— surprised because she feels so well. Her cardiologist told her that those who experience mild rejection soon after a heart transplant actually tend to do the best in the long run. But she takes nothing for granted. She has no doubt about her priorities. "I got what I wanted, which was to be here with my family," she says.

Stacie is one of approximately twelve hundred patients to be implanted with SynCardia's 70 cc artificial heart. Although the large heart is approved by the FDA as a bridge to transplant, SynCardia has now created a 50 cc heart, suitable for women and adolescents, and has obtained FDA approval to explore the use of the smaller heart as a destination therapy, meaning that the implant would be permanent. No one yet knows how long the TAH can last, but the technology is evolving at an incredible rate. The trend toward miniaturization continues, and a 30 cc artificial heart is now in development that performs just as well as the larger versions. In spite of the overwhelming success of the TAH, there are a few drawbacks. Patients must have an open incision with two tubes the size of small garden hoses extending from their abdomen to the driver. Then there is the need to recharge the driver on a daily basis, plus the inconvenience of carrying the backpack at all times. All of these issues pale in comparison when the alternative is death, but the technology is moving quickly toward a more advanced end product: a small, completely self-enclosed artificial heart with a long-life battery implanted under the skin, which eliminates the portable driver and obviates the need for an open incision.

The potential market for artificial hearts is huge. Heart disease is the number one killer of both men and women, and biological hearts are extremely scarce. Many people today die of heart disease when they are otherwise healthy. With a durable, dependable artificial heart, who knows how long their lives could be extended?

To obtain a physician's perspective on the artificial heart, I spoke with the pioneering heart-lung transplant surgeon Mark Plunkett, MD, who implanted three TAHs as chief of cardiothoracic surgery at the University of Kentucky (UK) College of Medicine. To speak with Dr. Plunkett is to hear the passion he has for artificial hearts and the incredible lifesaving alternative they offer to the "sickest of the sick" patients who would otherwise die waiting for a donor heart. While at UK, Dr. Plunkett pushed to have the medical center invest nearly $1 million to become equipped to offer the TAH to patients. A specialist in pediatric transplants, he has since accepted a position at the University of Florida and the Congenital Heart Center at Shands Children's Hospital in Gainesville, Florida.

Dr. Plunkett knew from earliest childhood what he wanted to do with his life. Growing up in a small town on Maryland's eastern shore, he was only seven or eight years old when he started telling people that he wanted to be a surgeon. He remembers vividly how, in 1967, when he was only seven years old, the world's first human-to-human heart transplant was performed by South African doctor Christiaan Barnard. The recipient, fifty-three-year-old Louis Washkansky, lived for only eighteen days until complications took his life. The transplant, however, was successful, and from that moment forward Plunkett knew that he wanted to be a transplant surgeon. He was already fascinated by medicine, and hearing about the transplant was "just over the top." He consumed medical information wherever he could find it and dedicated himself to clearing all the academic hurdles to becoming a surgeon. The farther along he got in his studies, the better he felt about his choice, and after an internship at Duke University Medical Center and a fellowship at UCLA Medical Center, he began performing heart and lung transplants on people of all ages, including children.

Plunkett was attracted to artificial organ transplant surgery in part because of the extreme shortage of biological organs, a problem that continues to plague patients with end-stage organ failure. There are over 119,000 people currently on waiting lists for donor organs, and about 7,000 of them die each year still waiting.[1] The problem is that, for various reasons, too few people opt to become organ donors. Meanwhile, with the growth and aging of the U.S. population, the need for organ transplants is growing. Additionally, the average heart transplant only lasts twelve to fifteen years before rejection and heart failure necessitate another transplant. This may not be such a problem in elderly people who may die of other causes before their heart fails, but for pediatric patients, it means that children must go through multiple heart transplants in a lifetime, and each time their transplant gives out, they are faced with the dire shortage of compatible hearts.

Several different artificial organs are now being created and tested, but the technology behind the artificial heart has long been in development. In 1963, Paul Winchell created the first artificial heart, which became the prototype for the Jarvik heart, first implanted in a human in 1983. Over the next few decades, artificial hearts became more sophisticated and patients who received them were living longer and longer. In Dr. Plunkett's words, the biotech industry is "exploding with gadgets," and doctors have to look carefully for the ones that prove safe and effective. He is excited about the SynCardia Total Artificial Heart, especially about the future prospects of smaller, totally enclosed hearts that could be a lifetime therapy for children and adolescents, a prospect that he believes is just around the corner. He is especially impressed by the TAH because it doesn't require drugs that suppress the immune system. The transplantation of a biological organ always requires immune system suppression, which only works for a while; in every case, the body will eventually reject the heart. He sees the next breakthrough in artificial hearts to be a miniaturized version with all the pumping power of the SynCardia heart, but only requiring the patient to wear a belt with a battery pack that continuously recharges the battery rather than the heavy backpack. The next version, a completely self-enclosed heart with a subcutaneous battery, is all but inevitable.

Getting a heart transplant is a complicated business on many levels. Having an artificial heart adds an entire layer of emotions that no one could have predicted twenty years ago. Dr. Plunkett described for me the coordination of the whole team of health care professionals who orchestrate the implantation of an artificial heart, and he described the process that takes sick and dying patients from the doctor's office to the operating table.

First the patient has to meet all the parameters that would make him or her eligible for a biological transplant. That means end-stage heart failure with biventricular failure, or the failure of both ventricles, that doesn't respond to any other implantable device or medication. The patient is generally referred to the transplant surgeon by a cardiologist, often a heart failure cardiologist. These patients are gravely ill, with very limited life expectancies if they don't receive a transplant. Their heart failure can be caused by a number of things, including coronary artery disease, heart attack, high blood pressure, viruses that attack the heart, or a disease of one or more of the valves. Generally the heart is grossly enlarged because as the heart muscle steadily weakens, it works harder and harder, causing it to enlarge, as would any muscle. While this temporarily improves the heart's pumping power, in the long run it makes the heart less and less efficient, a process that ultimately ends in death if no transplant takes place.

The symptoms of heart failure grow increasingly worse as the weakened pumping of the heart fails to provide adequate circulation to the vital organs, which begin to shut down. The patient feels overwhelming fatigue, even when at rest, and suffers shortness of breath as the blood and other fluids accumulate in the lungs, heart palpitations as the heart speeds up in its struggle to pump blood to the body, and swelling of the body due to the retention of fluids. Loss of appetite and nausea can also complicate the general feeling of sickness experienced due to the insufficient perfusion, or blood flow, to vital organs. Patients in this stage of heart failure are unable to tolerate any physical activity without extreme discomfort, and due to the fluid buildup in their lungs, they can feel like they're drowning. Many can sleep only when sitting up, a situation that exacerbates the already extreme fatigue. It is usually when they are in this condition that they are referred to a transplant surgeon like Dr. Plunkett.

Once it's determined that the patient is a candidate for an artificial heart, it takes a team of people—cardiologists, nurses or nurse practitioners, transplant coordinators, social workers, a psychologist, a nutritionist, an infectious disease specialist, and sometimes an ethicist—to manage all the facets of the case. These professionals meet on a regular basis to address all kinds of issues that patients face, including a lack of insurance or inadequate social support. The patient receives psychological counseling to help him understand and cope with the finality of removing his native heart and to address his anxieties and the flood of emotions that engulfs a person in what could ultimately prove to be the end of his life.

Dr. Plunkett said that when patients are told about the Total Artificial Heart and shown a short film about it, not everyone agrees to it. "It's just such a radical notion," he says. "A lot of older people have a hard time accepting it." People have understandably complicated feelings about the process. The heart has historically played an important role in our culture, in ways that go far beyond its biological function. For most of human history, the cessation of the heartbeat has been the very definition of death—so to be without a heart altogether is a terrifying prospect.

And, for thousands of years, the heart was thought of as the seat of our emotions, the source of love, the deepest feeling part of us. When we are really sure about something, we "know it in our heart." Without a biological heart, are we really "alive" in every sense of the word? Then there are questions that deal with fate and the lifelong expectation that once the heart gives out, one isn't "supposed" to be alive. For the religious patient, perhaps God meant for us to die at the specific moment that our heart gives out, and "cheating fate" comes with some type of eventual punishment. Dr. Plunkett has received all kinds of questions from patients who were introduced to the artificial heart. One patient asked him, "Will I still be able to fall in love if I don't have a heart?" For others, the thought of having an artificial pneumatic device driven by an external power source in place of a natural heart is simply too macabre. As exciting as the technology is, our society hasn't fully defined what will constitute humanity when our most vital function is carried out by a machine.

Still, nothing focuses the mind more than the knowledge of impending death, and most of the eligible patients who have been offered the TAH have gratefully accepted it. Younger patients are more receptive to having an artificial heart, and Dr. Plunkett has heard of European patients who have developed so much confidence in their artificial heart that even when a biological heart becomes available, they choose not to take it. When I asked Stacie Sumandig whether she would accept a permanent artificial heart if she had the option, she didn't hesitate to say yes. The TAH is a gift that bestows both quantity and quality of life to those who would otherwise die, and as time goes on, the TAH, or another version of an artificial heart, will promise good health and a greatly extended life span to those who receive it.

But there's another issue that has yet to be resolved, and that is how to know when to turn the device off. Dr. Plunkett described to me a hypothetical case in which this question would be front and center. Suppose that a patient with an artificial heart has a severe stroke that renders her brain-dead, but just like a heart-lung bypass machine, an artificial heart keeps the blood circulating throughout her body. The patient remains pink and warm, as though she were simply in a peaceful sleep. The family would be counseled that continued circulation support is futile because once all brain activity ceases, the person they know will never return. The patient is, for all intents and purposes, dead. It's likely that the medical team would recommend turning off the artificial heart—a decision similar to removing a ventilator from a brain-dead patient, but therein lies an agonizing decision.

Patients removed from ventilators do not usually die instantly. They can live for hours or days while their bodies slowly shut down, but the deactivation of an artificial heart would entail immediate death. If the proper medications are not administered prior to deactivation, there can be spasms and gasping breaths, as the body reacts to the sudden withdrawal of oxygen. This is only a reflex caused by a still-functioning brain stem, not a sign of returning life, but it can be misleading. No matter how appropriate it is to turn off the device, families may perceive their loved one as struggling for life and feel intense guilt. Their last memories with their loved one may haunt them forever. Some families will not do what they consider an act of murder, and no one knows how long a brain-dead person with an artificial heart would continue to have some of the biological functions that can make them seem alive. The artificial heart has not had its Karen Ann Quinlan or Terri Schiavo, but such a case is bound to occur. Although artificial heart transplants are too new for there to be a body of research about how such ethical questions can be resolved, we can get some idea by how end-of-life issues play out with some of the older implants that assist the heart.

In recent decades, heart patients have had an expanding menu of implantable devices. The oldest and most well-known cardiac device is the pacemaker; however, cardiac implants are being invented that can address specific problems short of end-stage biventricular heart failure. More and more people are receiving these devices, sometimes as a bridge to a heart transplant, but often as a permanent therapy. These devices include a biventricular assist device, which supports the function of both ventricles; an extracorporeal membrane oxygenator (ECMO), which can take over the function of the heart and lungs while a patient recovers from an acute event; an intraortic balloon pump, which can take over up to 20 percent of the heart's workload; a left-ventricle assist device; a right-ventricle assist device; mechanical circulatory support (a mechanical pump that circulates blood in a steady rather than a pulsing action); artificial valves; and an implantable cardioverter defibrillator (ICD), a device that delivers a powerful shock to "reset" the heart when an abnormal rhythm is detected. Former vice president Dick Cheney famously had an ICD before he later received a heart transplant.

The bioethicist Lynn A. Jansen has written about the deactivation of pacemakers, which generate steady electrical charges to keep the heart beating at a regular pace, when a patient is dying. For many patients, the pacemaker is a blessing that can greatly enhance and extend life when the heart's electrical pulses become dangerously irregular, putting them at risk of sudden death by cardiac arrest. However, now that many of the recipients of pacemakers are in old age and facing the end of life, often to causes other than heart problems, a functioning pacemaker can prolong the dying process—and the suffering—long after the patient would have naturally died. All too often, families are faced with the decision of whether and when to turn off the pacemaker and let natural death run its course. It is well recognized that patients with pacemakers develop a complex psychological relationship to their devices: they are both grateful to it and dependent on it, perhaps seeing it as the last bulwark between them and death. Some people actually see the pacemaker, once implanted, as part of the body, and it is within the boundaries of what we consider to be part of our body and what we consider to be not our body that a bioethical debate is raging.

Traditionally, the body is considered to be an essential part of the self, if not the entire self. Jansen has written that before we can make any decision about whether and when to deactivate a pacemaker, we have to decide whether it is or is not a fully integrated part of the body (and therefore of the self).[2] While the pacemaker clearly doesn't consist of human cells and tissues and originates from outside the body, once it is implanted, it becomes an active part of a system that is life sustaining. She presents the problem of a patient who is in a persistent vegetative state who would not be alive were it not for a functioning pacemaker. In Jansen's estimation, if the pacemaker is deemed to be part of the body, turning it off would be analogous to murder, whereas if the pacemaker is something that is considered extrinsic to the body, it can ethically be deactivated without the act being considered a killing. Jansen's position is that the pacemaker is not designed to function by itself, but only as part of a living system. If that system has shut down, the pacemaker loses its meaning and intention. Just having blood circulating in a body that is brain-dead does not constitute life.

The bioethical debate has picked up lately as the recipients of implantable cardioverter defibrillators (ICDs) face the end of life. ICDs essentially do the same thing that the external defi-

brillating paddles do to prevent sudden cardiac death by shocking the heart back into a life-sustaining rhythm. Only the ICDs
are internal implants, and while they are capable of performing
numerous functions for a failing heart (including pacing), they
are best known for being able to detect an arrhythmia and immediately deliver a powerful shock that "resets" the heart and
staves off death. Patients who have received one of these shocks
describe it as a powerful, disturbing blow that feels like "being
kicked in the chest by a horse." Although patients feel dependent
on the ICD for life, most develop a sense of anxiety about when
they are going to receive one of these painful shocks, and after
having received one, it is disturbing to know how close they
were to death. Patients feel both attached to their ICDs and afraid
of them at the same time.

As effective as the ICDs are when a too-rapid or irregular
heartbeat occurs, they do not prevent death from advanced heart
failure. Designed to prevent sudden cardiac death, they can subject patients to a much worse kind of death, delivering shock
after shock to a heart that is slowing or becoming irregular near
the end of life. The process is agonizing for dying patients and
traumatic for their loved ones. The body receiving the shocks can
be violently jolted as the patient dies. There have been instances
reported wherein a dying patient received thirty or more shocks
while their loved ones watched helplessly, when they otherwise
might have slipped away peacefully.

There are several reasons why more and more patients are
suffering such an agonizing death. This is one area where human
nature, and our traditional attitudes about the dying process,
have not caught up with advanced technology. The clear solution
is to deactivate the ICD's defibrillating function when the patient
has just days or weeks to live. Turning off the ICD does not mean

that sudden death will occur. The heart will continue to beat until natural death happens, yet there seems to be a kind of conspiracy of silence among doctors, patients, and family members about deactivating ICDs, and discussions about this almost never happen.

Studies have shown that patients are reluctant to discuss turning off their device because of the complicated emotions they feel about it, and often liken deactivation to suicide. They almost universally do not want to make the decision about when to deactivate their ICD, looking instead to their doctors to do so. However, the few studies that exist show that doctors don't want the responsibility of deciding, either, and rarely discuss the option of deactivation with their patients. They look to their patients to initiate the discussion about when to deactivate, and up to one-third of doctors will never deactivate an ICD because they consider doing so physician-assisted suicide. The manufacturers of the ICD can send a technician to the patient to turn off the device wirelessly, but one of the three American makers of the ICD insists that any decision to deactivate has to be between the patient and the physician. "We don't think it's our job to practice medicine," an industry representative told *The Washington Post* in 2006.[3] So, by default, too often no decision is made, and what could otherwise be a peaceful death turns into a tortuous process for everyone involved.

First introduced in the 1980s, ICDs have been implanted in more than 500,000 Americans, and each year about 150,000 new ones are implanted. With the aging of the baby boomers and their parents, many of those patients are approaching the end of their life. In some cases, these patients have long ago lost touch with the physician who implanted their ICD, and find that no one among their current caregivers even wants to have a conversation about deactivating the device. A strong bias exists among patients, doctors, and nurses, and even hospice care providers don't like to discuss deactivation of an ICD. Harvard Medical School reports that even though most hospices have powerful magnets that, when held over the ICD, can stop the shocks, only 10 percent of hospices have a policy regarding the issue.[4]

A 2011 article in the *American Journal of Nursing* by James E. Russo is one of the most often-cited commentaries on the realities of ICD use at the end of life. Russo conducted a survey of articles published from January 1, 1999, through October 31, 2010, on the subject, including surveys of patients, physicians, industry representatives, and hospice workers. His article sheds light on why, in the vast majority of cases, ICDs are not deactivated when a patient is dying. Russo notes that defibrillators are used primarily to treat a too-fast or chaotic heartbeat involving the heart's ventricles, the lower chambers of the heart. The devices are usually implanted many years before a patient is considered to be near death, and doctors generally do not see any urgency to discuss deactivation at the time of implantation. In addition, they come to be regarded as an integrated part of the body, a distinction not given to other forms of life support such as a ventilator or a dialysis machine. Compounding these factors is the psychological relationship that patients have with their ICDs. Russo found evidence that patients often overestimate the lifesaving ability of the ICD, considering the device "a trusted friend."[5] Patients may mistakenly think that the ICD can prevent death from advanced heart failure, but this is not the case. Russo also notes that patients who have received shocks from their ICD can develop anxiety and even depression due to both the fear of the shock and the fear of death. There is a kind of Catch-22 cycle that patients are subject to, inhibiting them from ever discussing the possibility of deactivation with their doctors.

One small study by Nathan Goldstein and his colleagues and

published in the *Journal of General Internal Medicine* in 2008 describes the attitudes of fifteen patients who were assembled as a kind of focus group to examine their attitudes about deactivation. It should be noted that none of these patients was actively dying, so the idea of deactivation seemed more hypothetical than urgent to them. Amazingly, none of the patients had ever discussed deactivation with their physicians, and they were uniformly reluctant to discuss the topic, even though they also feared being shocked by their ICD. They tended to overestimate both the benefits of the ICD and the risk of immediate death should the device be turned off. None of them even knew that deactivation or reprogramming (to turn off the defibrillating function while continuing the pacing function) was even a possibility.[6] One patient likened the deactivation of an ICD to "an act of suicide."

The study also showed that the patients didn't really understand the role their ICD played in their health or why they had received it. All of them had a high level of anxiety about receiving shocks, whether they had already experienced one or not. Not one of the participants was willing to discuss deactivation either for the study or at a later date with their physicians. One equated deactivation of an ICD to sudden cardiac arrest, an inaccurate notion but one that was strongly held throughout the group. On the other hand, the participants did not seem to understand that the ICD would not prevent them from death due to advancing heart failure or another disease. Most "could not contemplate any situation in which death was a likely or probable outcome," and several explicitly said that it should be the physician's job to recommend deactivation at the right time. The dependence of patients on their doctors to make the critical decision for them is highly problematic, because another study by some of the same authors found that doctors are no more willing to discuss deactivation than are patients.

To explore the physician's point of view, Goldstein and his team conducted detailed interviews with four electrophysiologists, four cardiologists, and four general practitioners. The results of this study are probably a bit skewed because the authors included only those physicians who had at some time discussed deactivation with a patient (a decided minority of doctors). Even so, these physicians very rarely had that conversation with a patient, and they identified several impediments to initiating such a discussion. Chief among their reasons was a fear of making a patient feel that he or she is dying and that they are "shutting off the hope."[7] The doctors also acknowledged that, even when making decisions at the end of life, there is something about the ICD that makes it inherently more difficult to deactivate than other life-support technologies. They reported the fact that the ICD is small, implanted in the body, and easy to overlook, and they also admitted that they were not comfortable discussing it with the patient. While all of the doctors thought the discussion *should* take place, they were unsure of when to initiate it and had strong reservations about any mention of the possibility of impending death.

Although the authors don't mention it explicitly, the implied running theme is the tendency throughout American medicine to avoid recognizing the inevitability of death. Both patients and physicians are loath to admit that there is not always another treatment or intervention yet to be tried that can stave off death. Not only are physicians reluctant to discuss deactivation of an ICD at the end of life, it follows that they are at least as reluctant to actually deactivate the devices. Russo cites a study showing that out of a sampling of electrophysiologists and device company representatives (who often end up deactivating the device), only 57 percent said they would feel comfortable turning off an ICD even after a patient requested it.[8] So we find ourselves in a conundrum: patients want their doctors to make the decision for them while doctors put the onus on patients, and a large percentage of industry representatives are not comfortable deactivating ICDs even when the patient asks for it. The result is that it is extraordinarily rare for an ICD to be deactivated when death is near, and an increasing number of individuals are suffering what almost anyone would call a "bad death." Because doctors (and patients) tend to equate deactivation with euthanasia, they sometimes delay the decision even more by ordering ethical or psychiatric evaluations, increasing the chances that a patient will experience painful and futile shocks.

As we look ahead to emerging medical technologies, it's likely that the unintended consequences created by ICDs are likely to occur with a range of products. Even when discussions about eventually deactivating an ICD occur at the time of implantation, such early discussions will not necessarily reflect a patient's actual feelings when he has days or weeks to live. Nor will they necessarily reflect what loved ones might decide under duress when a patient is comatose or brain-dead. The fact that even physicians show a different attitude toward a small, implanted electronic device that works in concert with the body's natural functions shows that the line between the patient and the technology is already blurring. If we believe that these devices are all that stand between us and death, we are bound to be deeply emotionally attached to them and to reflexively resist efforts to disengage them.

As the science of artificial hearts and cardiac-assist devices rapidly advances, even more radical solutions are being investigated. In a few decades, even the totally self-contained artificial heart could be obsolete as scientists devise a nanotech solution to a failing heart. In one possible scenario, millions of nanobots could be programmed to perform the functions of the heart, oxygenating the blood cells throughout the body and removing and eliminating carbon dioxide and other waste materials. The bots could even be programmed for self-propulsion, meaning they would propel themselves through the bloodstream, ending the need for a heart altogether. Such a miracle cure could add an unknown number of years or decades to life, but once again, the technology will eventually come up against the ultimate limit of the human life span. There will most likely come a time when, as patients enter extreme old age and suffer other, catastrophic health problems, the nanobots could keep them alive in a state of great suffering. The scientists working on such problems will have the ability to deactivate the nanobots, but here we find ourselves in the familiar moral quandary as to when to "turn off" the bots.

But let's consider the tangible possibilities that are most likely in store for those of us alive today. It is quite reasonable to assume that many people will end up with some type of implanted device that performs some vitally important bodily function for them, whether it's an artificial organ, one of a plethora of cardiac-assist devices, or even a neural implant. As we become habituated to these devices, they create a "new normal" for us, and our sense of identity likely expands to include our devices. That new normal, for many people, will mean that their life cannot go on for long without the help of the implant. We're only beginning to get used to the idea of turning off external life-support machines when a patient is brain-dead; to deactivate an internal "part of the body" feels even more like killing, and asking to have an implant turned off seems even more like suicide. While we rush to embrace the life-giving and life-extending benefits of artificial implants, we are setting ourselves (and our loved ones) up for some very difficult decisions later on.

There have been efforts in recent years to sort out the issues of deactivation and to establish some guidelines for doctors who are faced with the requests of terminally ill patients. The American Heart Association, along with the American College of Cardiology and the Heart Rhythm Society, in 2008 issued their Guidelines for Device-Based Therapy of Cardiac Rhythm Abnormalities. A research team led by Richard Zellner examined these guidelines for a paper that was published in 2009 in the journal *Circulation: Arrhythmia and Electrophysiology*. They note that even though withdrawing a life-sustaining treatment "goes against the grain" of physician training and values, the guidelines make it clear that deactivation of a pacemaker or ICD in a dying patient "should not be regarded as either physician-assisted suicide or euthanasia."[9] The guidelines emphasize the principle of patient autonomy, a central concept in modern medical ethics. That principle asserts that doctors should above all honor patients' wishes, and if they object to doing so on moral grounds, they must refer the patient to another physician who can do so. The authors also draw upon the writings of the physician and Franciscan friar Daniel P. Sulmasy, who wrote the highly influential article, "Within You/Without You: Biotechnology, Ontology, and Ethics."

Sulmasy argues that there is a clear ethical distinction between medical technologies that constitute "replacement" therapy (such as a *biological* heart transplant, which replaces the function of the patient's native heart) and a "substitute" therapy (such as a pacemaker, which does nothing to cure the underlying disease, but imposes an *artificial* functionality as long as it is turned on). Sulmasy clearly identifies artificial implants as substitute therapies that can ethically be deactivated upon a dying patient's request. He doesn't make a distinction, as doctors and patients clearly do, as to whether the technology exists inside the body, like an implanted device, or external to it, such as a dialysis machine. Substitute therapies, to Sulmasy, are all artificial interventions that do not cure an underlying condition. The rules pertaining to discontinuing any artificial life-support therapy in a dying patient are all the same, whether that therapy is an artificial ventilator or an implanted electronic device.[10] He notes that the acceptability of discontinuing life support when a dying patient asks for it or when the patient's pain or discomfort outweighs the benefits of the technology have by now been well established. The problem is that in the real world, neither doctors nor patients are likely to have read Sulmasy's articles or to appreciate the nuances he describes. He admits that, as Goldstein et al. demonstrated above, "what seems equivalent according to the logic of ethics continues to feel psychologically different to both patients and practitioners."[11]

Sulmasy's view, while eminently reasonable, makes a major distinction between artificial implants, which can be ethically shut off under certain conditions, and biological implants, with which it would be unethical to interfere. It may not be so clear to everyone why a biological implant is not an artificial technology whereas an electronic implant is, and he does not grapple with the issue of an aged or permanently disabled person who is not actively dying requesting the deactivation of her implant because she simply wishes to die. And as of now, only a few years later, medical technology is already poised to eclipse Sulmasy's organic/inorganic distinction.

To further challenge the biological-artificial distinction, scientists are now working on engineered organs that integrate artificial parts with human biological cells, as I will discuss in the next chapter. Implantable devices are being designed that will become far more integrated into the body and are harder to classify as either organic or inorganic. In addition to that, not everyone is psychologically capable of the kind of dispassionate view of implantable devices that bioethicists exercise. People have deep attachments to both their own and others' bodies, fear death, and are sometimes emotionally incapable of deciding to turn off life-support technologies for an incapacitated loved one. Even highly educated physicians have complicated emotions when it comes to "giving up" on a dying patient and admitting "failure." It's not likely that deeply entrenched psychological and cultural phenomena are going to change at the same pace at which science is moving today.

It's unavoidable that a time will come when death is upon us, and our implants could greatly complicate our passing. It's not likely that the decisions for when to deactivate implants will get easier in the foreseeable future. Even today, doctors find it very difficult to predict just how much life a gravely ill person has left, and if we don't initiate conversations with them about deactivation, it's unlikely they will ever take place. But the greater conversation, about how to understand our "selves" as partly biological and partly artificial, has barely just begun.

# 3

## The Race to Beat Kidney, Lung, and Liver Disease

To look at Frank Bowers, one would never guess the struggle he has had with chronic kidney disease. A very young-looking sixty-four-year-old father of three adult children, Frank spends three and a half hours, three times a week, hooked up to a dialysis machine. His natural kidney function is down to 5 percent, and without the dialysis, he would die from the buildup of toxins that his kidneys can no longer filter from his blood. Frank has end stage renal disease, or ESRD. As an African-American, he is four times more likely to develop ESRD than the general population, and he comes from a family in which kidney disease has devastated the lives of two of his siblings and a cousin. He worries for his three children and twelve grandchildren, and it's largely for their benefit that he's participating in a worldwide study on genetics and kidney disease.

Frank has always been a stickler for leading a healthy, active lifestyle. Even so, at the age of forty-eight, his kidneys started to fail. There was no pain, but he felt dizzy, disoriented, and listless, so his doctor ordered a kidney workup. Protein showed up in his urine, a sure sign that his kidneys were failing, but after three biopsies and test after test, his doctors were never able to find out what was causing the problem. They prescribed a special diet, increasing his water intake, and treatment with the steroid drug prednisone. These changes helped for a few years, but in 1996, his kidney function collapsed. His doctor prescribed dialysis and Frank was put on a transplant list.

He waited thirty-seven months for a new kidney, and during those months, his health deteriorated terribly. He almost completely stopped urinating and, as a result, started to retain massive amounts of fluid. His doctors assured him that once he received a new kidney, he would be a new man. But first they had to find a genetic match—something thousands of people die waiting for each year. Luckily, a match was found, and Frank underwent transplantation surgery.

The benefits of the new kidney started while Frank was still in surgery. When the surgeon attached the last blood vessel to the kidney, he started to urinate immediately. When he woke up, he says, "I felt twenty years younger. I worked, went on vacations, and was able to hike and exercise. It truly felt like a miracle." For the next thirteen years, Frank savored every day, but eventually he came up against the limits of organ transplantation. The new kidney failed, and he became gravely ill again.

By January 2013, his fluid retention was so extreme that he weighed 456 pounds, an experience he describes as "miserable." The water that his kidneys were unable to expel literally doubled his body weight. He was admitted to the hospital and told that his condition was terminal. Family members were called in to say their last good-byes. But his doctors weren't ready to give up on him, and despite the dire prognosis, they placed a catheter in his groin to facilitate more dialysis. To everyone's surprise, the dialysis helped more than anticipated. He was immensely relieved when he quickly lost massive amounts of fluid and returned to his normal body weight of 226.

Since then, Frank has spent three days a week hooked up to a dialysis machine. The dialysis drains him of energy and after each session, all he can do is go home and crawl into bed. He has good days and bad days, when the lethargy is overwhelming, and he knows he's still living on borrowed time. Another complication that often affects people on dialysis has become a major issue. Doctors have to implant a graft in the blood vessels of dialysis patients that allows their blood to flow out through a tube, be filtered by the dialysis machine to remove toxins and excess fluid, then returned to the body. Frank has now had multiple grafts implanted, and each one has worked only for a while before getting clogged or failing due to the collapse of the blood vessel used as access. After so many years of dialysis, doctors are running out of blood vessels to work with. Frank knows only too well that the day will come when he has no more viable blood vessels that can accommodate a catheter.

On the cold November day when I met with Frank in his cozy townhouse, beautifully decorated by his wife, I couldn't help thinking how unfair it was for a man so obviously full of life to be in such a terrible predicament. It was clear that his children and grandchildren were the focal point of his life, and he worried about each of them developing kidney failure. I hesitated to ask him what comes next. Clearly, another kidney transplant could give him another decade or so of life, but was he interested in going through the ordeal of the surgery, plus the immune system suppression and all the medical management that accompanies it?

"I haven't talked to my doctor yet about another transplant," he admitted. In addition to ESRD, he now has congestive heart failure, and he might not survive the surgery. He's also concerned about what the immune system suppression would mean for him as he got older. He was at the point where every available choice carried potentially deadly risks and may not even provide an acceptable quality of life. "What about an artificial kidney?" I asked him. "Would you be willing to try one if you had access to it?" He thought about it for a minute and said, "Okay, if that option existed, I would be willing to be a guinea pig for my children and grandchildren."

As mentioned above, African-Americans are far more likely to develop end-stage kidney disease than other ethnic groups. Although the reasons are not well understood, it's clear that the problem involves genetics and major risk factors such as diabetes.[1] Most of the patients referred for dialysis have had chronic kidney disease for many years and their natural kidney function is down to 10–15 percent. As Frank's story illustrates, dialysis prolongs life, but it is time-consuming, has horrible side effects, and carries the ever-present risk of complications due to the repeated failure of multiple grafts over time. In addition, while dialysis buys the patient some time, the underlying kidney disease will continue to progress, and, without an organ transplant, the disease will eventually be fatal.

Even with a transplant, it's a little-known fact that no organ transplant is likely to last for the duration of the patient's natural life. All transplanted organs either become subject to the disease that destroyed the original organ or are eventually rejected by the body. The immune system can be suppressed for a time, but it will always win the war against even genetically matched organs. Meanwhile, the demand for organs is rapidly growing in the United States due to the aging of the population, and we are in the midst of a dire organ shortage.

According to the United Network for Organ Sharing (UNOS), which works with the Department of Health and Human Services to operate the Organ Procurement and Transplantation

Network, thousands of Americans die each year while waiting for a genetically matched organ. There are currently about 120,000 people on the transplant waiting list.[2] About 90,000 of them are on the list for a kidney, yet only about 18,000 kidney transplants are performed each year.[3] The problem is that far too few people elect to donate their organs. Why? The problem is due to both a lack of information and some common myths that deter people from becoming donors.

One of the most common myths is the belief that if emergency room doctors know you're an organ donor, they won't work as hard as possible to save your life. According to UNOS, this idea goes completely against the grain for medical professionals, whose first priority is to save a life. The death of a patient for doctors and nurses is regarded as a failure, no matter what the circumstances, and medical personnel suffer high rates of emotional trauma, and sometimes professional stigma, when patients die. In addition, having volunteered to become an organ donor is not enough in many states to complete the process when death is approaching—one's next of kin must also agree to the donation, and can easily take that option off the table if they suspect that their loved one is not receiving the lifesaving care he or she needs. The few studies that have been done on physicians' attitudes about prolonging life in dying patients have shown that doctors are in fact biased toward prolonging life as much as possible even when heroic measures go against patients' stated wishes. At least in the United States, the most common scenario is one in which doctors prolong life far beyond the point when that life can have any meaning.

Other reasons why people don't elect to become organ donors include the erroneous belief that their religion forbids it, the notion that they are too old to donate (there is no limit to donation even among very old people), and the distressing idea that organ donation mutilates the body and makes it unfit for an open-casket funeral. In reality, organs being surgically removed has no effect on the appearance of the body, and in no way makes it inappropriate for an open-casket funeral. Still, the very thought of violating the integrity of the body can be so upsetting to loved ones that they will not give permission even when they know there is no logical reason to refuse it. When I was in my early twenties, I learned that one of my brothers had decided to become an organ donor when renewing his driver's license. I was so shocked by the idea that I tearfully begged him to reverse his decision. Knowing what I know today about the life-giving process organ donation is, I still find it easier to be an organ donor myself than to think of my loved ones donating.

According to the World Health Organization, the shortage of organs is a global problem that has led to a thriving and highly exploitative black market in organ trafficking.[4] The practice is fed by demand in economically developed nations like the United States and the desperate plight of poor people in developing nations. One major ring of organ smugglers was apprehended in 2004 for purchasing kidneys from live donors in poor South American countries and selling them to recipients in Israel, the United States, and Europe. The ring was run out of Israel, which has one of the lowest rates of organ donation, due to the misconception that the Jewish religion forbids it.

The business of organ trafficking is extremely lucrative. A 2004 article in *The New York Times* reported that live kidney donors in Brazil, for example, can make six to ten thousand dollars, a fortune in a country where the minimum wage is only eighty dollars per month.[5] The process is fairly elaborate, entailing the recruitment of donors who are flown to South Africa to have their kidneys removed, then returned to their home country, while the traffickers charge up to $150,000 for one kidney. The fact that it's illegal to pay for an organ in the United States has done nothing to stop the organ recipients, who are up against impending death, from paying whatever price the traffickers demand. They fly to South Africa to receive their new organs and, after a period of observation, return to the United States. South Africa has recently overtaken Turkey as the country of choice for the surgical procedures, even though the trade in organs is technically illegal. Obviously, for these procedures to take place, many hospitals, surgeons, and other medical personnel must look the other way.

Even if many more people elected to donate their organs, the practice of organ transplantation has a natural limit: the challenge of genetic matching and the eventual rejection of transplanted organs. Just as in Frank's case, all transplants will eventually be overtaken by the underlying disease that necessitated them or they will eventually be rejected by the body. This is especially problematic for people who receive organ transplants at a young age. Those people will need multiple transplants across their lifetime, and due to the above-mentioned limitations, such a prospect is by no means guaranteed. And even in the best of circumstances, with voluntary donation, there are still ethical problems that must be surmounted before organs can be donated.

Even when a donor who is dying has been identified, medical teams face quandaries such as the choice to administer heparin, a blood thinner that prevents blood clots in the organs, but could hasten the death of a bleeding patient. Another advance in preserving organs in the dying or brain-dead patient is the introduction of extracorporeal interval support for organ retrieval (EISOR). This is the temporary use of the ECMO machine that was used by Stacie Sumandig after her artificial heart transplant. EISOR uses the ECMO machine to remove and oxygenate the blood from the body, then return it to the body to perfuse and preserve organs prior to removal. The reason this is needed is because once the donor's breathing and blood circulation cease, organ damage starts to occur very quickly, lessening if not ruining the chance of having transplantable organs. When EISOR is used, steps must be taken to prevent the reanimation of a heart that has stopped. The heart will start beating again due to the blood flow and without the patient breathing, something that can give grieving loved ones false hope that the person could be revived, even when brain death has occurred, and the blood circulation will immediately cease as soon as the EISOR is discontinued. One can never underestimate the difficulty of allowing the removal of a loved one's organs when the body appears pink and is warm to the touch. Families often feel that they would be "killing" the donor even when brain death has been confirmed by multiple tests. Not everyone understands or is emotionally prepared to accept that brain death is death, and that the artificial maintenance of blood circulation will do nothing to reverse that fact.

In addition to the above issues, which pertain to all cases of organ donation, even the issue of prescribing dialysis entails ethical problems that must be addressed. In 2002, writing in the *American Society for Artificial Internal Organs Journal,* Eli Friedman listed some of the following questions that doctors must already ask when considering whether to place a patient on dialysis or a transplant list: Must dialysis be initiated in patients with a futile prognosis, and should advanced age exclude one from dialysis? Should the age of the recipient be a factor in prioritizing the allocation of organs? Should a child-to-parent kidney transplant be performed? Is heroin or crack abuse sufficient reason to prevent dialysis? Should a wealthy hospital contributor be given priority for a donor kidney? Should kidneys be marketed, meaning the rich buy organs from the poor? Is abusive, destructive, and noncompliant behavior reason for exclusion from dialysis?[6]

The use of permanent artificial organs surmounts some of these difficulties, and scientists throughout the world are now working on artificial organs that can be completely self-contained (meaning no open wounds for tubes or wires) for virtually every organ in the body. As Stacie's story illustrates, we are close to the day when there will be a permanent implantable organ for a whole range of deadly diseases. The effects of this phenomenon could mean a dramatic extension of life span for the recipients of these organs (or devices), and a dramatic lessening of the difficulties entailed in natural organ donation. Artificial organs are not subject to hindrance through some of the above issues, but they still pose issues when it comes to what bioethicists call distributive justice. The shortage of natural organs certainly invokes the problem of deciding who will receive natural organs, but artificial ones, which are extremely expensive to develop and distribute, may only be available to those who live in wealthy nations, have insurance that covers them, or have the ability to pay outright, leaving the poor and the middle class out in the cold.

Artificial organs will soon be designed that are more durable and perhaps more powerful than natural ones, leading them to become not only curative but enhancing. Among bioethicists, genetic enhancement, for example, is widely regarded as unacceptable, but will that thinking remain when (a) genetic enhancements start to be delivered in the service of correcting a disease or deficiency, and (b) artificial implants can go beyond correction to the ability to enhance human attributes? We should consider that we already perform a huge number of medical procedures each year that have no medical benefit and simply provide aesthetic enhancement in the form of cosmetic surgery. Plastic surgery was originally developed as a way to correct disfigurement from accidents and disease, but millions of people have now sought it out as a way to improve their appearance. Likewise, drugs that were originally developed to treat narcolepsy are now in use by the armed forces to boost wakefulness and concentration in our troops. It's reasonable to predict that technology that is now in development to treat a range of diseases and disabilities will be sought after as ways to improve "normal" functioning.

Even when the above-mentioned ethical questions are answered satisfactorily and the patient is put on dialysis, dialysis is by no means a perfect technology. Only about 33 percent of patients on dialysis survive for five years, whereas 80 percent of transplant patients have a five-year survival rate.[7] Dialysis doesn't replace all of the functions of a natural kidney, such as helping to regulate blood pressure, producing vitamins and hormones, lowering acid levels in the blood, and allowing the needed reabsorption of liquids back into the body. Patients on dialysis, like Frank, still experience the sickness associated with the buildup of toxins in their blood, which is only relieved three times a week, as opposed to the continuous filtration provided by the kidneys.

Because of these limitations, there are several research teams working on the development of an artificial kidney that would perform more of the necessary functions on a continuous basis. Most of the artificial kidneys under development entail having an external device that can be worn on a belt, but they also include tubes and wires that must be connected to a patient's bloodstream through an open wound—a recipe for potentially deadly infections and the blood vessel collapse that plagues patients on dialysis. Another concern is one of social acceptance. Stacie Sumandig's story illustrates that we have a long way to go before everyone is comfortable seeing and interacting with a person who has an artificial external device performing a life-sustaining function.

Some of the most exciting research for kidney replacement is being conducted by an interdisciplinary team of bioengineers, biologists, medical doctors, and other researchers at the University of California, San Francisco (UCSF), called The Kidney Project. Headed by the bioengineer Dr. Shuvo Roy, this project is an example of the convergence of several leading-edge technologies. The kidney being developed by this team combines microelectromechanical science (the miniaturization of electromechanical devices), nanotechnology (the use of superstrong materials built by almost infinitesimally small molecules), and human cell technologies. A summary of the technology can be found on UCSF's Web site, which says, "Advances in silicon nanotechnology were required to make it possible to mass-produce reliable, high-porosity, robust, and compact membranes. Improvements in molecular coatings that impart blood compatibility and techniques to coat silicon membranes, without blocking pores, were needed. And cell sourcing and storage issues had to be resolved. All of these technologies are now in place."[8]

The device under development is implantable, freestanding, and powered by the patient's own blood pressure. It will require no batteries or external tubes and would provide far more of the kidney's functions than dialysis. About the size of a coffee cup, this two-part device includes actual kidney cells sourced from stem cells in the patient's immune system and, unlike the implantation of a biological kidney, would not require suppression of the immune system. The artificial kidney has already been successfully tested in rats, sheep, and pigs, and Dr. Roy predicts that human clinical trials will begin in mid-2017.

In addition to providing the 24/7 toxin-reducing blood filtration close to that of a biological kidney, the artificial kidney promises to be far cheaper to manage than dialysis. Each dialysis patient now costs Medicare about eighty-five thousand dollars per year. The management of a transplant recipient costs Medicare about thirty thousand dollars per year, mostly for the antirejection drugs. The artificial kidney is projected to cost ten to twenty thousand dollars per year after the initial surgery to implant it.[9] This dramatic drop in price will, without doubt, be exceeded by further cost reductions as the technology matures and becomes widely available, while it is a given that further refinement of the technology will result in greater miniaturization. Since the artificial kidney has no battery, patients will not need repeated surgeries to replace it, as is currently the case for pacemakers and other implantable devices. And as it becomes smaller, it will eventually be placed just beneath the skin so that various components can be replaced over the life of the device without major surgery.

In the artificial kidney, the two major components include a hemofilter and a cell bioreactor. The filter uses a nanotech coating that is so finely grained that sugar and salt molecules and other toxins that build up in the blood can be filtered out while blood clots can be prevented and important molecules such as proteins can stay in the blood. The bioreactor, which contains real human cells taken from the patient, will further process the filtered blood so that the appropriate amounts of sugar and salt are returned to the blood and water is reabsorbed into the body, creating urine that can be expelled by the bladder. So far, both components have been successfully tested in animals, and the research team is working on combining these components into one small device that is attached to the patient's internal blood vessels. Although it's too early to know how long each component will last once implanted into the body, this device is not being developed as a bridge to transplant—it is intended to be a permanent solution to ESRD.

The first human clinical trials will focus on the safety of the device and will be tried in patients who are on the transplant list but strong enough to survive the surgery—not the sickest patients or those who have been on the list for a very long time. As advanced as the technology is, getting it to work is not the biggest challenge—obtaining sufficient funding is the biggest "if" in this scenario. Assuming that the project can attract the needed funding, Dr. Roy estimates that the clinical trials will be completed by 2020, after which the device will become available to the public.[10]

This project, although it promises to be a lifesaving gift to patients with ESRD, raises many interesting ethical questions that go beyond the considerations that play into choices of how to prioritize recipients for a biological kidney. For example, does the presence of human cells in an apparatus that is joined to the human circulatory system and thus becomes integrated with the body's natural processes (such as blood composition and blood pressure) mean that the device is "alive"? Is it "human"? Furthermore, does control of the device rest with the recipient, the doctors, or even with society? Consider whether the recipient has the right to have the implant deactivated or removed when this action would quickly lead to death. In the case of an artificial kidney, a person suffering terribly near the end of life from another disease or condition might prefer the relatively peaceful death brought about by kidney failure to a painful death by the other condition. Would this decision be regarded as suicide? In addition, if deactivating or removing the device entails the assistance of medical professionals, would their acts be defined as euthanasia or even murder? Once again, the state of our science has exceeded the current boundaries of medical ethics and social, legal, and political policy. The artificial kidney represents one of the first forays into a blurring of the boundaries between a living system and a purely technological one.

Funding for The Kidney Project, almost $7 million to date, has so far come from some unexpected sources—at least I was surprised to learn that it has been funded not only by the National Institutes of Health (NIH), but also by the National Aeronautics and Space Administration (NASA) and the Department of Defense (DoD). While funding is still needed from private investors to bring the project to full fruition, the investment from NASA and the DoD suggests that the technology being developed could have broad applications to a number of major innovations, many of which go beyond the boundaries of medicine.

The Kidney Project is not the first major initiative involving funding from sources that might be considered strange bedfellows. The very first serious examination of the implications of converging technologies took place in the United States in 2001 in a study that was funded by the National Science Foundation (NSF), the military, and the Department of Commerce.[11] Both the research and the larger debate about CTs spans many countries, including the United States, Germany, and other countries of the European Union and Asia. The mix of funding sources shows how CTs have the potential to affect almost every aspect of life. In 2003, a 482-page report funded by the NSF and the DoD, "Converging Technologies for Improving Human Performance: Nanotechnology, Biotechnology, Information Technology and

Cognitive Science," was released (the NBIC report). In it, editors Mihail C. Roco and William Sims Bainbridge write:

> Beyond the 20-year time span, or outside the boundaries of high technology, convergence can have significant impacts in such areas as work efficiency, the human body and mind throughout the life cycle, communication and education, mental health, aeronautics and space flight, food and farming, sustainable and intelligent environments, self-presentation and fashion, and transformation of civilization.[12]

The report reflects the massive effort on the part of many individuals and institutions to coordinate the work of specialists across a wide range of disciplines. Included in these disciplines are not only scientists and doctors, but also ethicists, politicians, and prominent thinkers throughout the humanities. This shows a recognition that traditional demarcations between the physical and biological sciences and the humanities cannot hold in a world of radical technological transformation where the need for keeping human dignity at the center of innovation is more urgent than ever. Scientists must learn to speak the language of ethicists, religionists, philosophers, sociologists, and politicians while these groups must become more conversant in CT breakthroughs. Major changes are needed in nearly every sector of society as a new paradigm asserts itself.

CTs will mean unprecedented cooperation across a wider range of specialties than society or its teaching institutions have ever supported. In the NBIC report mentioned above, Roco and Bainbridge lay out their recommendations as follows: Individual scientists and engineers must gain skills in all neighboring disciplines to enhance collaboration. In academia, institutions must "undertake major curricular and organizational reforms to restructure the teaching and research of science and engineering." The federal government must place a priority on funding "converging technologies focused on enhancing human performance, including research on the social and ethical aspects of CT." Professional societies should work to "reduce the barriers that inhibit individuals from working across disciplines," and the press bears the responsibility of providing "high-quality coverage of science and technology, on the basis of the new convergent paradigm," so that citizens can participate intelligently in the formulation of public policy. This is a tall order, which will entail unprecedented change. These new synergies will challenge all of society to be smarter, to encompass a broad range of divergent knowledge, and to adapt to radical changes that have never been encountered before. The move toward CTs will demand that the extreme specialization of education and work models, which has held sway for over a century, will no longer be possible.

The use of powerful new technologies such as nanotechnology and artificial intelligence, when combined, means not just incremental change, but change that is exponential. Perhaps the most difficult concept to grasp is one at the very essence of CTs, which is spelled out in an article for *Nanotechnology Spotlight*. The author states that "learning from nature and from imitating it in order to create artifacts is increasingly turning into a construction of new bridges between the living and non-living, or a modification of natural processes and structures for the purpose of design, which even extend to the vision of technically creating biological entities from scratch."[13]

The article explains the situation this way: "New brain-machine interfaces, prostheses to compensate for sensory limitations and improve one's motor ability, and also visions about implants that can improve cognitive achievement all belong to the core topics of the CT debate, especially with regard to the topic of human enhancement."[14] Not only will we exercise new concerns about entirely new biological-artificial life forms, we will *be* these life forms. A burning question remains—will we have the foresight and adaptability to properly manage our new cyborg nature as we integrate more and more of these powerful innovations, or will we be taken by surprise by potentially disastrous uses of CT?

Scientists say that, as daunting as the scientific and ethical issues are in the creation of artificial organs, the real trick is obtaining the necessary funding stream that will move a technology from the lab to animal trials, then to human trials. The process of obtaining FDA approval can take up to ten years in a very complicated regulatory environment, and new technologies need continuous funding to cross one hurdle after another. Since artificial organs are being developed at a number of biotech firms that are subject to the vicissitudes of the marketplace, the funding can dry up before the company has cleared all of the requirements for obtaining FDA approval.

A case in point is the HepatAssist artificial liver, whose checkered history is confusing and disheartening to those of us who aren't immersed in the biotechnology universe. To start with, the liver is a far more complex organ and performs many more functions than the kidney. It breaks down toxic substances absorbed from the intestines and other parts of the body and synthesizes them into harmless molecules that end up being excreted from the body. The liver also plays a key role in digestion, converting food into proteins, carbohydrates, and fats, including cholesterol. We know that too much cholesterol is harmful, but it is essential in the creation of important hormones such as estrogen, testosterone, and adrenal hormones. The liver also holds reserves of sugar and releases it into the bloodstream when the body needs it, and converts food into important compounds used in the creation of blood-clotting factors.

The technical difficulties of creating an artificial liver are enormous, but many of these difficulties have recently been resolved. Achilles Demetriou, a former surgeon at Cedars-Sinai Medical Center in Los Angeles, in the early 2000s led a team of researchers and doctors who created an artificial liver support system called the HepatAssist. Like the artificial kidney, the HepatAssist combines artificial components with living cells—in this case porcine hepatocytes, or liver cells from pigs. The device was created as an extracorporeal (outside the body) bridge to transplant or a support system for liver healing in patients with severe liver failure. One of the functions of the technology is to separate the plasma from red blood cells in the blood so that the plasma can be circulated through the various components, filtered through the porcine cells and other parts, then returned to the bloodstream minus many of the toxic substances that build up during liver failure. The device draws upon nanotechnology to create a filtering membrane with pores so tiny that cellular debris can be prevented from being returned to the patient's bloodstream.

In 2004, Dr. Demetriou's team published the results of a study involving a sizeable group of patients with liver failure who received support from the artificial liver, and compared them to results in those who received no support. The researchers cautioned, and many experts agreed, that the results were somewhat skewed because of the inclusion of patients who had previously received liver transplants and rejected them, a group for whom no treatments have been known to work, but the study nevertheless showed that patients who received support from the HepatAssist had a significantly higher survival rate at thirty days.[15] Experts who subsequently analyzed the results agreed that the patient survival rate would have been markedly greater if no patients who had rejected transplants had been included, but the study underscored the promise of the technology to serve as a life-extending bridge to transplant. However, the HepatAssist was in for a maze of difficulties in obtaining the needed funding to further develop the technology.

Before the study was published, the firm that had developed some of the technology, Circe Biomedical, Inc., in Lexington, Massachusetts, bought the rights to the HepatAssist technology. Circe had already made breakthroughs in raising pigs that were free of viruses and whose cells could be safely used in the device. The FDA approved the HepatAssist for further study in acutely ill patients, but to realistically gain full approval for medical use, Circe needed to do additional research excluding patients who had rejected transplants to obtain more accurate results. The researchers already knew that the device helped those who had viral or drug-induced liver damage, but widespread trials involving many more patients and multiple research centers needed to be done, which is a very expensive process.

Circe was in the process of applying for permission to do the additional studies when the 9/11 terrorist attacks rocked the financial markets, and the venture capital that would have funded the study dried up. Circe sold the technology to another biotech firm called Arbios. Arbios tried to resurrect the trial but could not attract the funding, so it sold the technology rights to another firm, HepaLife Technologies, and there it seemed that the story was in danger of ending. Dr. Philip Rosenthal, one of the authors of the original study, speculated that the firm either couldn't raise enough money to continue the new research or decided that the device would not be profitable and dropped the project. Dr. Rosenthal believes that since the HepatAssist was developed, interest in the medical community was shifting to the possibility of using human stem cells to grow specialized cells in the lab to replace damaged tissue in diseased livers.

The HepatAssist seemed out of luck, but the story only grew more convoluted. After acquiring the HepatAssist technology, HepaLife renamed the device the HepaMate. In a 2009 statement, it announced that the HepaMate technology, when acquired, was FDA-approved for further research and enjoyed fast-track status as well as clinical data from phase I/II and phase II/III clinical trials. They stated that they expected to initiate new phase III clinical trials, but did not say when. The same document, which is aimed at potential investors, makes it abundantly clear that no profits had been made from the HepaMate technology, and there was no guarantee that any profits would ever be made—boilerplate language for legal protection. They cite a list of caveats that pose challenges to all biotech companies seeking to take a new technology from the research bench to the bedside.

Among the many challenges presented, they say, "The biotechnology industry is characterized by intense competition, rapid product development and technological change. A number of companies, research institutions and universities are working on technologies and products that may be similar and/or potentially competitive with our cell-based artificial liver." In other words, a competitor could develop a better device and bring it to market first, rendering the HepaMate obsolete, and there is no guarantee that the device will meet with regulatory approval. In addition, competition among companies for talented researchers and other personnel is also intense, and there is no certainty that these individuals will be recruited and retained. And those are just the challenges to product development; a whole new set of issues comes into play when the product is ready to be marketed. Marketing issues include the need to establish brand recognition, developing distribution networks and relationships with insurance companies, health care professionals, and patients, and the ability to offer rebates and discounts to make the product competitive, plus the need to settle any legal disputes.

The parent company of HepaLife, a firm called Alliqua Biomedical, had created HepaLife to develop its liver therapies, and continued to work on porcine embryonic stem cells, which it developed into hepatocytes (liver cells), for use in the HepaMate. In February 2009, HepaLife announced that it was planning the needed phase III trials that would not include transplant patients who had rejected new livers. In December 2010, HepaLife changed its name to Alliqua Biomedical, Inc. It's not clear what happened with the plans for phase III trials, but a Web site offering investor information in December 2012 announced once again that HepaLife (now Alliqua) was still planning phase III trials. The next news came in February 2013, when Alliqua announced that it was "strengthening and realigning" its executive team. I called Alliqua several times in search of someone who could give me the status of the HepaMate, but no one at the firm returned my calls. It appeared that my search was at a dead end when I tried to contact Dr. Demetriou, the lead author of the pivotal 2004 article, and found that he had passed away in 2013. The most recent information I could find was a notice, in November 2014, that the Web site domain name, HepaLife.com, was for sale.

One thing in regard to the artificial liver is clear: so far none of its many owners has been able to gain enough traction to bring this technology even close to market. It's evident that attracting and maintaining the huge amounts of money needed to complete this tortuous journey is very difficult. It was disheartening to think that such an important and beneficial device, which could eventually lead to a permanent, implantable liver, was so dependent on the vicissitudes of the marketplace to become widely available to those whose lives hang in the balance. But that is what often happens when a new medical technology is developed in the private sector.

Dr. Robert Bartlett was already a trailblazer when he took on the project of creating an artificial lung that would provide 100 percent of a patient's oxygen needs. A professor and surgeon at the University of Michigan, he is the inventor of the ECMO machine, as long ago as 1966. The ECMO, although a lifesaving device, has certain limitations. It can only be used for a maximum of eight to ten weeks, ironically, because it does all of the work of the heart and the lungs. When the heart and lungs are taken offline, deconditioning takes place rapidly, and they can be dangerously weakened as a result. Recently, Bartlett has been working with other surgeons, biomedical engineers, pulmonary specialists, and a transplant team on the BioLung, an artificial lung that is powered by the patient's natural heartbeat and allows him or her to be ambulatory while waiting for a transplant. The BioLung can also be used to give damaged lungs time to heal in the case of smoke inhalation or some other trauma, where a transplant is not an option.

The lungs rival the heart in being essential to the maintenance of life. Ridding the blood of carbon dioxide, a normal byproduct of metabolism, and exchanging it with oxygen is critical to maintaining every cell, tissue, and organ. Even brief deprivations of oxygen can be catastrophic to the body and especially to the brain. In his 1993 bestselling book, *How We Die,* Sherwin Nuland writes, "If one were to name the universal factor in all death, whether cellular or planetary, it would certainly be loss of oxygen." He goes on to quote Dr. Milton Helpern as saying, "Death may be due to a wide variety of diseases and disorders, but in every case the underlying physiological cause is a breakdown in the body's oxygen cycle."[16]

Under normal conditions, the critical function of our lungs is something we never have to think about; breathing is controlled by the autonomic nervous system, which receives its marching orders from specialized cells in the brain stem. The lungs work in concert with the heart to exchange carbon dioxide with oxygen and to ensure that oxygen makes it to every cell in the body to support the myriad processes of metabolism. This gas exchange occurs continuously as deoxygenated blood is moved from the heart through the lungs, where it is oxygenated, returned to the heart, and then recirculated throughout the body.

Diseases of the lungs that lead to disability and death generally cause a loss in the size, shape, and elasticity of the intricate airways of the lung. These include cancer, chronic obstructive pulmonary disease (COPD), and cystic fibrosis. Transplantable lungs are in short supply, and over 200,000 Americans die of lung disease each year. About 50,000 of those people are on a transplant list but don't survive long enough for genetically matched lungs to become available.

In 2001, Dr. Bartlett and Dr. Keith Cook constructed the first version of the BioLung. It was a major advance in that it required no power source other than the patient's heartbeat, and this allowed the heart to remain active and fit while obviating the need for a battery that would require periodic replacement. The current model, which is being developed by the medical device company Michigan Critical Care Consultants (MC3), is entirely composed of lightweight polymers, is self-enclosed (meaning the blood never leaves the body), and is only about the size of a soda can. It has been extensively tested in animals and is awaiting FDA approval for human trials.

The BioLung is implanted into the patient's chest and the

heart pumps blood into it via an implanted catheter. The blood filters through a bundle of hollow plastic fibers with nano-sized holes so tiny that only gas molecules can pass through them. It is through these fibers that carbon dioxide is exchanged with oxygen. The device can be programmed to send the oxygenated blood back to the heart or it can pass through the natural lungs for additional filtration before being pumped throughout the body.

The BioLung has several advantages over the ECMO machine. It provides 100 percent of the patient's oxygen needs, requires no open wound or power supply, and frees the patient from being tethered to an external machine. In addition, it doesn't necessitate the removal of the patient's lungs, so people with severe lung damage can eventually revert back to their natural lungs once healing has taken place. With all of these advantages, the BioLung is not yet designed to be permanent, but it is estimated that it could last up to five years.[17]

I spoke with Dr. Bartlett in 2013, on the eve of the commencement of human trials. At that time, an external, wearable version had been tested on one hundred patients worldwide. He predicted that within the next ten years, there will be a permanent implantable BioLung that will address the dire lack of transplantable organs. The BioLung is not the only artificial lung under development, and with the rapid advance of technologies such as nanotechnology, biomedical engineering, and stem cell technologies (to explore the possibility of growing biological lungs), we may soon have a permanent, completely self-contained artificial lung that could outlast any biological transplant. While artificial lungs are subject to certain risks, such as the risk of blood clots, the durable polymers being used in them would not be subject to the ravages of disease and aging. Some version of the artificial lung may integrate both biological and artificial components, much like the artificial kidney. A person whose overall health is reasonably good but who suffers from lung disease could add decades to her life.

One extremely important aspect of the BioLung is that, ethically speaking, there would be no need for deactivation at the end of life. If the lung is powered by the patient's heartbeat, when the heart ceases to beat, the lung would naturally cease to function. This may not be true of all artificial lungs that make it to the marketplace, but it is the type of issue that biomedical engineers and other researchers need to think through at the inception of creating such devices. An artificial lung that would require deactivation in order to cease the steady oxygenation of blood would invariably run afoul of medical ethics at the end of life.

It's likely that artificial organs will be among the first advanced technologies to be widely accepted. The reason is that in every individual, parts of the body age or deteriorate at different rates, depending on genetics, environment, and lifestyle choices. An obvious example is that of a smoker whose lungs deteriorate much faster than any of his other organs, or a patient with inherited kidney disease. As in the case with Stacie Sumandig, artificial organs are likely to be accepted during crisis situations when the life of an otherwise healthy person is at stake and there is no biological organ available for transplant. As time goes by, artificial organs can be replaced if they wear out or are simply rendered obsolete by technical innovations that surpass them. Another advantage of organs made completely from artificial materials is that they don't pose the risk of rejection that is a factor in all biological organ transplants. It's feasible that in the near future, aging individuals could accumulate several artificial organs, tissues, joints, and other devices, so that the person contains multiple artificial components that his or her life depends on. But let's consider what this would mean if scientists don't conquer the aging process itself.

While there are individual variations in how quickly the parts of the body age, aging is still an all-encompassing process that affects the entire body inside and out. It is due to a general winding down of the body's cellular processes, the bedrock of life itself. Our cells divide throughout our lives, and each division results in the copying of the original cell. Slowly and over time, minor mistakes creep in when copying the cellular DNA, which we might consider as the software that tells the cell what to do. Mistakes then get copied and additional mistakes take place. The effect accumulates over a lifetime, rendering our cells less efficient, slower to regenerate, and often diseased. Although we can replace various parts as they wear out and die, aging affects every cell of the body. As long as this process continues, people will face hard choices when considering artificial organs. Should a brand-new heart, for example, be implanted into the body of a very frail and sick older person whose quality of life is already severely compromised? What if that person has Alzheimer's disease or another form of dementia that science has yet to cure? Would extending such a life really be a compassionate decision?

In his book *Our Posthuman Future*, Francis Fukuyama presents a dark vision of the social aspects of extending life through artificial means. He writes:

> The only problem is that there are many subtle and not-so-subtle aspects of human aging that the biotech industry hasn't quite figured out how to fix: people grow mentally rigid and increasingly fixed in their views as

they age, and try as they might, they can't make themselves sexually attractive to each other and continue to
long for partners of reproductive age. Worst of all, they
just refuse to get out of the way, not just of their children,
but their grandchildren and great-grandchildren.[18]

While I don't agree with Fukuyama's vision in general, there
is no denying that using artificial organs to extend life will introduce the need for social and cultural adjustments. Human life
does not occur in a vacuum; it occurs within the context of an
interconnected community in a world of limited resources. I'll
discuss the issues of greatly extended life spans in greater detail
in chapter 7, but there is no doubt that if people live to be two
hundred, three hundred, or even older, human reproduction cannot continue at the current rate. More to the point—the aging
process will need to be arrested or radically slowed down in order for very long lives to be desirable or even to make sense.

The rapid advance of new biomedical technologies means
dramatically expanded control over how we live and die, yet it is
an open question whether people will welcome such control or
be equipped to deal with it. We place a very high value on human
life, but does valuing life extend to the point of prolonging it
when a continuing heartbeat means only prolonged suffering?
This is where we are today, with no real social or cultural mechanisms in place to guide people who are going through agonizing
decision processes for themselves or a loved one. Scholars disagree
on the advantages of maintaining life in the presence of extreme
suffering, but doctors tend to come down on the side of retaining
life at any cost, and this is already causing widespread, unnecessary suffering at the end of life. There seems to be a kind of
whistling-through-the-graveyard avoidance of this issue in which
neither doctors nor patients and their families want to take on

the profound responsibility of deciding how we die. The unavoidable fact, however, is that many of us alive today *will* decide how we die. One of the major impediments to allowing life to end without heroic (and futile) interventions is an outdated legal system that often criminalizes the compassionate act of allowing a person to die even when intervention would only postpone the inevitable and at the cost of great suffering.

As a lover of animals, I'm often struck by the difference in our attitudes toward the end of our pets' lives. Virtually no one questions the rightness of putting a suffering animal "to sleep," yet we steadfastly oppose withholding meaningless treatments from suffering humans. We are, in fact, deeply ambivalent about suffering. All of human history can be seen as one vast enterprise to ameliorate human suffering, whether it be the establishment of better health, more access to food and comfort, or the lessening of vulnerability to all the vicissitudes of life. Yet literature, philosophy, and theology are replete with accounts of meaning derived from human suffering. Many contemporary bioethicists, especially the bioconservatives, argue that human life can still have meaning even when great suffering is present, and that therefore some good can come from it.

Some may argue that the fact that people do often find (or make) meaning from suffering is only an artifact of a world in which we have no choice. There is no denying that much suffering has been an enduring aspect of life since the beginning of time. War, disease, bereavement, natural disasters—you name it—have plagued humankind from the beginning and continue to do so. But there is also no denying that the sweep of human history, despite many setbacks, has been in the direction of alleviating the common causes that have made life "nasty, brutish, and short." The emergence of science, as well as cultural evolution, is rapidly bringing about a world that is vastly improved over any that has existed before.

Not everyone sees the march of medical and other technologies as benign. This is especially true of those who believe that suffering is a critical component, a kind of refining fire that everyone must go through in order to become morally and spiritually mature—to be fully human. Fukuyama writes:

[T]he deepest fear that people express about technology is not a utilitarian one at all. It is rather a fear that, in the end, biotechnology will cause us in some way to lose our humanity—that is, some essential quality that has always underpinned our sense of who we are and where we are going, despite all of the evident changes that have taken place in the human condition through the course of history. Worse yet, we might make this change without recognizing that we had lost something of great value. We might thus emerge on the other side of a great divide between human and posthuman history and not even see that the watershed has been breached because we lost sight of what that essence was.[19]

Whether or not Fukuyama was considering the meaning of suffering when he wrote the above passage, it seems that his historical view of humanity necessarily includes the assumption of a life profoundly touched by suffering. However, mankind's constant striving to end suffering in ourselves and others can easily be seen as a hardwired trait because of its remarkable universality and perseverance. As long as suffering looms so large in human life, many thinkers will continue to place a moral value on it and will continue to argue that suffering is an integral component to human wisdom. But we are hardly able to see this issue objectively from our current perspective of vulnerability to a multitude of ills. Not until we find ourselves on another plateau, from the perspective of vastly improved and extended lives, will we be better able to sort out the meaning of suffering. And it is entirely possible that, even when so vastly improved, life will still entail great existential uncertainties that will provide ample opportunity for the famed character-building experience of suffering.

# 4

## Got Diabetes? There's an App for That

It was by no means a random occurrence that Michelle Craig became one of the first people to test-drive a cutting-edge new artificial pancreas. The tech-savvy Web content manager and mother of two teenagers in Roanoke, Virginia, has been dealing with type 1 diabetes, the most aggressive form of the disease, since the age of six. As a tech enthusiast, she had been doing her homework for years when it came to searching out the latest innovations for diabetes treatment. When I spoke with Michelle, what struck me most about her was her upbeat attitude and intelligent focus on searching out every possible way to better control the condition that overshadowed her life.

When she was a child, her parents tried to maintain as much normalcy as possible for Michelle and her sister, who was also diagnosed with juvenile diabetes at a very young age. Michelle was allowed only limited amounts of sugar and remembers when the awful-tasting Tab was the only sugar-free soft drink on the market. She rejoiced when Nutrasweet became available, slightly expanding the range of dietary choices for diabetics. By the time she reached junior high school, she had learned to partially control her blood sugar by pricking her fingers and giving herself insulin shots. She ran track and went to summer camp along with all the "normal" kids. It was only after she became an adult that her parents told her that as a child, some other parents had refused to let her come over for sleepovers because they feared what might happen if she ate the wrong thing or had a medical emergency. Diabetes runs in both sides of her family, and her father had the disease himself, so they knew full well what their daughters had to contend with.

In her twenties, after she married but before she had her first child, Michelle was given an insulin pump, which provided her body with a more steady release of insulin than the shots had. Before the pump, she had still been dealing with the blood sugar seesaw of highs and lows, and the pump was an improvement. It also made pregnancy safer, and she was able to have two children who so far do not have the disease. In 1997, when she got the pump, it was still the best solution for controlling severe diabetes. However, in 2000 she discovered that a number of artificial pancreases were under development that promised to dramatically improve the management of blood sugar. She followed the topic avidly and learned to search for human clinical trials on ClinicalTrials.gov, the government Web site listing all trials being conducted with NIH funding. When a trial came up for an artificial pancreas being developed at the University of Virginia (UVA), she jumped at the opportunity to apply as a trial subject. The Web site included contact information for a coordinator of the UVA project, and Michelle "begged to be included." A short time later, she was thrilled when she found out she had been accepted.

The goal of the trial was to have patients test out the safety and effectiveness of the technology in their own homes, so she had to make several trips, a one-and-a-half-hour drive, to Charlottesville to pick up the devices, download data, be trained on how to use the technology, and discuss other issues with the doc-tors overseeing the trial. For Michelle, this involved several phases. The artificial pancreas was in actuality the sum of three devices—a continuous glucose monitor (CGM), an insulin pump, and a smartphone. In July 2014, she was fitted with the CGM, a small device connected through a needle in her abdomen to her blood supply. Just as its name suggests, this device performed continuous readings of her blood glucose levels for the duration of the trial. About two and a half weeks later, she had her insu-lin pump upgraded to one that utilizes Bluetooth technology for wireless data transmission. Through Bluetooth, this device was able to communicate glucose levels to a cell phone that runs on Android software and that had a specially created app called the Diabetes Assistant. Her job was to enter key information such as what she had eaten, and how much, into the app. The app, which also received blood glucose readings from the CGM, then wire-lessly told the insulin pump how much insulin to release into her bloodstream.

Michelle used the full configuration for two and a half weeks, and she describes the experience as "a world of difference from just the insulin pump." During the trial, she says, "I could count on one hand the number of times my blood sugar was out-side of the normal range." She felt "very, very well," and best of all, she had no need to prick her fingers or give herself shots. The hardest part for her was "not to react to every blood sugar read-ing and let the devices do the work." After so many years of hav-ing to constantly check her blood sugar, it was a challenge just letting go of control. But it was also incredibly liberating. The phone had to be recharged every night like any other cell phone, and she had to refill the insulin in her pump every three days. I asked her if she had any anxiety about the phone going dead and not having access to a power supply. "Not at all," she replied.

"Because if the phone dies the insulin pump [which is connected to the continuous glucose monitor] will take over." As for the complexity of having three devices all talking to each other, she said, "If you can use a smartphone, you can do this."

I asked her if there was any downside to the technology she tested. "My only complaint is that I'd like to lose some of the devices," she said. "It would be great if something like the Apple watch could replace the smartphone." And she expressed hope that neither the FDA nor the insurance companies will put obstacles in the path of making the technology widely available. She recognized that perhaps the most important thing about the artificial pancreas is that, by better controlling glucose levels, it will greatly reduce the long-term, severe complications of organ damage, nerve damage, and vision loss that diabetes can entail. The technology, she said, will lead to "a lot more productive, happy people." If there is a permanent artificial pancreas in her future, she will gladly embrace it.

The UVA artificial pancreas technology is a perfect example of the kind of wide-ranging collaborations that are introducing technology to our bodies on a scale as never before. The project has included collaboration among endocrinologists, computer scientists and programmers, diabetes specialists, and engineers at several universities and research institutions.

Dr. Boris Kovatchev, who designed some of the algorithms used in the technology, has followed an unusual path to becoming UVA's director of the Center for Diabetes Technology. A native of Bulgaria, he received most of his training in mathematics, probability, and statistics at Sofia University. He arrived in the United States as a student in 1991 and soon found himself doing postdoctoral work at UVA on the problem of diabetes blood glucose control. His father had the disease, and he was interested in how he could apply mathematical modeling to help those who suffer from the type 1 version of the condition. Little did he know that fourteen years later, he would be collaborating with the range of specialists he has worked with while developing an artificial pancreas.

So what does the artificial pancreas have to do with mathematics? It uses specially designed algorithms in the smartphone app that Michelle tested out. Dr. Kovatchev's technology, which enjoys FDA fast-track status and is planned for another, longer set of clinical trials, liberates patients from a previous version of the artificial pancreas that connected them to a laptop computer via numerous movement-restricting wires. The laptops also ran on special algorithms, but they were clunky and had to be tested while the patient was in a hospital so that doctors and nurses could periodically check their glucose levels and administer insulin shots. The new version benefits not only from better algorithms, but from electronic miniaturization and wireless technology. What a difference a few years of innovation have made.

When I spoke with Kovatchev in February 2015, he told me that the UVA artificial pancreas is not perfect, but is likely one link in a chain of innovations that will result in better and improved models. The current version doesn't do everything that a natural pancreas does. Blood glucose levels can rise and fall at lightning-quick speed, and a biological organ is able to calibrate blood sugar levels more quickly. Patients with the artificial pancreas can still experience spikes and troughs that result in either organ-damaging hyperglycemia or dangerous episodes of hypoglycemia, when the blood sugar is too low. Hypoglycemia can cause a host of frightening effects from dizziness to coma and even death. The pancreatic hormone glucagon can raise low blood sugar, and there are other versions of the artificial pancreas that are attempting to administer both insulin and glucagon as needed to keep blood sugar as steady as possible. At the rate this technology is progressing, it's probably only a matter of a few years before artificial pancreases perform very near the level of a natural organ and become what Dr. Kovatchev calls "a mainstream solution" that will obviate the need for biological transplants.

Dr. Kovatchev said it is probable that within the next two to three years, the UVA artificial pancreas will be brought to market. The project has received ample funding from the National Institutes of Health and the Juvenile Diabetes Research Foundation. Near the end of our conversation, I mentioned The Kidney Project at UCSF and how it combines artificial materials and components with living kidney cells to bring about a closer approximation to a biological kidney, and I asked him whether a similar approach is being tried in an artificial pancreas. The short answer is yes, and there are several lines of research that are exploring the use of both human and animal cells that will be combined with hardware to better control blood sugar. While this is a very promising concept on a scientific level, it would open up the resulting devices to the same ethical quandaries as The Kidney Project. It's not clear where the living cells would come from—the patient's own stem cells or donor cells, which is controversial if the cells are taken from human embryos. On the other hand, if the cells come from animals (pigs are sometimes used as the source of transplantable tissues), there would be the issue of immune system suppression and possible animal viruses.

So far we've considered the artificial heart, lungs, kidney, liver, and pancreas. But research is rapidly advancing in creating ever more sophisticated devices to replace virtually every part of the body, including organs like the skin and even tissues such as light-sensing artificial retinas. It's reasonable to assume that in the next few decades, bionic technology will be able to replace virtually any part of the body. And, arguably, what technology can replace, it can improve upon. Let's consider the issues of implanting hardware that works with wireless data transmission.

At some point in the near future, most artificial organs will be able to communicate vital data about the body via wireless transmission to doctors and specialists who can interpret that data for better management of diseases and conditions. This is, for the most part, a highly beneficial technology. I happen to have such a device myself. A few years ago, I had a gastric pacer, a small device similar to a heart pacemaker, implanted and connected via electrical leads to my stomach. I have a condition called gastroparesis, which simply means paralysis of the stomach. This stops the necessary contractions needed to break down food and even move fluids through the digestive system. The condition causes pain and swelling, and the worst part of it is incapacitating nausea. Before I received my pacer, the nausea attacks became so frequent and severe that I feared I would become completely disabled.

Fortunately, I lived in the Washington, D.C., metro area and had access to the most cutting-edge technology. My gastroenterologist referred me to the gastrointestinal surgeon Dr. Frederick Brody, at George Washington University Hospital. Dr. Brody was then one of the few surgeons in the country implanting the new devices into patients like me. By the time I got to Dr. Brody, I had lost forty-five pounds in one year and was so sick that my postoperative healing process seemed to take forever. After the surgery to place all the leads and implant the pacer, I wondered if I would ever be able to make it up a flight of stairs in my house again. It took several weeks, but slowly my stomach started responding to the electrical signals coming from the pacer. The nausea receded and soon I was able to eat solid food again. The strength and energy started to flow back into me, and it was only then that I realized how weak I had been before the pacer. The pacer has not, and does not, completely cure the disease, but it has given me about 80 percent of my life back, something for which I'm deeply grateful.

Today I only need to see Dr. Brody about four times a year. During our visits, he is able to place a wireless sensor across my stomach that reads how much electrical juice is going to my stomach and transmits that information to a boxlike receiver. He can not only check on the frequency and strength of the pulses coming from the pacer, but also determine how much charge remains in the battery. If I'm having problems, he can wirelessly adjust the pulses, generally making them stronger or longer-lasting. When the battery is spent, the whole pacer must be replaced—a minor surgical procedure now that all the leads are in. My first pacer lasted two years before it had to be replaced, and for whatever reason, the second pacer seems to be working even better than the first one. It's completely enclosed beneath the skin in my abdomen. There is no open incision and the only maintenance needed is the quarterly visits with a knowledgeable surgeon like Dr. Brody. It may sound naïve, but the pacer really opened my eyes to how much the body is an electrochemical phenomenon—a technology, in a sense, that can integrate other forms of technology and even work in concert with them. The body doesn't seem to care whether this technology is artificial or biological, as long as the necessary functions are being performed. In fact, the integration of "hardware" can greatly enhance the performance of "wetware" in the body.

I don't know whether the gastric pacer will extend my life (I suspect it will), but I know it has helped me to avoid a great deal of suffering and disability. That being the case, I have a certain amount of anxiety about what could happen if my digestive system stops responding to it. As Dr. Brody has told me, there has to be some natural function of the body for the device to work. The stomach could lose all ability to contract and then the pacer would have no significant effect. I try not to think about this scenario, which, after all, may never happen. Still, being a bionic being myself has opened my mind to the amazing promise and the profound risks of our growing dependence on medical technology.

Of all the devices considered here, one capability they all have is wireless computing technology. As time goes by and more people are fitted with more devices, the question will arise as to who "owns" the vast amount of data being transmitted to doctors and technicians. Do we own our own data, or does our doctor? Who will have a right to see this data? The hospitals and vast medical conglomerates that run most of them today? Our family members who might need to make informed medical decisions for us should we become incapacitated? Insurance companies making decisions about whether to sell us coverage? Employers considering whether to hire us? What about the government and other entities who might gather and maintain databases designed to track health statistics? Could our data be sold the same way mailing lists are sold to direct-mail fund-raisers and other organizations who hope to sell us their products or services? In short, will the existence of this data open us up to new forms of discrimination and privacy invasion that we have yet to even think of?

As bionic devices and wearable computing—clothes and accessories that perform continuous readings of one's heart rate, blood pressure, etc.—create a gigantic database of biological information about us, where will this data be stored? Most concerning of all, will we have the legal and technical ability to protect the data from being hacked? (These same issues also have bearing on people who have had their genomes sequenced.) Given the extent to which encrypted financial data has been hacked in the last few years, we can assume that our health information is just as vulnerable. Even with comprehensive privacy laws, should they be uniformly enacted, it's doubtful that we will ever have complete electronic privacy.

The question is even more pressing considering that doctors, hospitals, and virtually all health organizations are rapidly moving toward electronic health records (EHRs). Typed records are replacing the notoriously bad handwriting of physicians that can be hard to read and can lead to medical mistakes—sometimes with fatal consequences. EHRs can contain your medical history, inoculations, existing conditions, lab and radiology test results, medications you're on, family history, vital signs, physician's notes, and just about any type of health information that you can think of, consolidated so that any doctor you see can access your entire health history. These records can more easily be shared with specialists who handle some part of a patient's care without the need to decipher handwritten notes. And, in the event of an emergency, a doctor can send vital information to a hospital while a patient is en route via ambulance, preparing emergency room personnel for her arrival even if she's unconscious and unable to communicate.

The combination of accuracy, efficiency, and convenience, not to mention the saving of lives, makes EHRs inevitable throughout medicine in the near future. But we return to the same essential questions: Who owns EHRs and with whom can they be shared? Can they be bought and sold? We know they can be hacked, but what regulations are in place to protect the privacy and security of our health information by law-abiding entities who nevertheless want to legally game the system?

So far EHRs are simply regulated the same way as any other medical record under the Health Insurance Portability and

Accountability Act of 1996 (HIPAA) Privacy Rule. HIPAA requires doctors to utilize certain protections of your EHRs such as the use of passwords and PIN numbers, data encryption, and an "audit trail" that keeps a record of who has accessed your records, what changes were made, and when. However, no matter how diligent health care providers are about adhering to these standards, clever hackers can always get around them. HIPAA requires hospitals to notify patients when a security breach has occurred, but this does nothing to correct the situation once a breach has already happened. You can assume that if you have an implanted medical device feeding vital information to your doctor, it becomes part of your EHR and is subject to the above benefits and risks. I'll return to the uses of EHRs presently, but first let's take a look at the U.S. military's role as a major, but little-known catalyst for the development of advanced medical therapies based on CTs.

# 5

## Jump-Started by the U.S. Military

As beneficial as devices created through converging technologies are in the medical sphere, these powerful convergences will have staggering effects when applied to warfare. As mentioned previously, the Department of Defense and the Pentagon are pouring billions of dollars into research and development (R&D) on converging technologies. The results are already taking us into new territory that only a few years ago seemed like science fiction.

Many of the same technologies being used in medicine are now being used to help develop weapons of war and what we might call supersoldiers. Unmanned drones, night-vision goggles, and drugs that boost alertness are already in use, but in the very near future, warfare will look dramatically different than it does today. Soldiers will be enhanced beyond recognition, and much more ground warfare will be conducted by robots. Human soldiers will become less and less directly engaged in fighting, and remote fighting will go to a new level. Today unmanned drones still require a human decision maker to strike a target, but soon combat robots will make such decisions. Many people find this possibility terrifying. I recently asked a friend, "What's more frightening, a human that kills or a robot that kills?" Without hesitation, he immediately answered, "A robot that kills. A robot has no conscience." I countered that while this is true, there are humans that actually *enjoy* killing. It's well known that humans are capable of terrible things when they can act anonymously and do not have to confront their victims face-to-face. On the other hand, machines can't be reasoned with and one can't appeal to their emotions. They do what they are programmed to do, and even "learning" machines that can incorporate new information have a limited range of behaviors. I can't help feeling that we both have a valid point, but there *is* something uniquely terrifying about a killing machine.

The military is deep into research on technologies that integrate biology, chemistry, physics, engineering, electronics, and materials science to create nanotech products such as extremely small sensors for the body and machines, superstrong fabrics that can resist bullets while absorbing almost no heat, and particles that can destroy biological and chemical agents without harm to soldiers. War is a terrible business, but out of the horrors of war have come some of our most commonplace technologies. This is especially true in the area of military medicine, where the intense pressures of the battlefield have given birth to some of the most significant breakthroughs.

The need for speed and mobile medical units have to a great extent created modern medical trauma care as we know it. The concept of triage was developed by the military to help prioritize the most seriously wounded soldiers when large numbers of them flowed into treatment centers following a battle. Millions of people can now thank the military for their face-lifts, since numerous plastic surgery procedures were developed in the military sector to reconstruct the faces and bodies of soldiers disfigured by injuries. Even though penicillin was discovered in 1928, it only became mass produced during World War II to treat infected wounds received on the battlefield. The technique of freezing blood and then slowly thawing it to limit cell damage—commonly known as blood banking—is a military innovation that also took hold during WWII. The first wound adhesives were developed and first used during the Vietnam War.[1] In addition to these breakthroughs, the Global Positioning System (GPS), freeze-drying perishable foods and medicines, programmable computers, and microwave ovens are all examples of military technology that transitioned to civilian use.[2]

According to the National Science Foundation/Department of Commerce report "Converging Technologies for Improving Human Performance," the U.S. military is now working on some of the most cutting-edge medical science using converging technologies. Better armor has saved lives but left many surviving soldiers needing more artificial limbs, so the military is developing some of the most advanced prosthetics, including powered knee and ankle-foot prostheses that produce a more natural gait. Other military technologies either in development or already being used include portable X-ray systems that can be taken into the battle zone and moved whenever and wherever they are needed; an FDA-approved blood warmer used to resuscitate soldiers with both severe hemorrhage and hypothermia; a motorized exoskeleton that will help soldiers move more quickly over any kind of terrain while carrying much more, and heavier, equipment and supplies; prosthetic limbs that work with propriocep-tors (special receptors in tissues such as muscles, tendons, and joints) to form a feedback loop allowing the patient to "feel" with artificial limbs; nanostructured coatings to provide ultrasmooth surfaces in implants, reducing the friction of metal-to-metal surfaces so that devices last longer; and a new cell-delivery system to speed up wound healing. This technique takes stem cells from fatty tissue and uses them as tiny cell factories to produce large numbers of new cells that are then injected into wound sites. This is now being explored as a way to heal damaged cartilage—something that could have breakthrough implications for the treatment of arthritis. These technologies, or those that succeed, will have no problem crossing over into civilian use. In fact, there is already a large market for them throughout the world.

Considering just some of the above technologies, it's not hard to envision lifesaving scenarios for civilians. Imagine a driver skidding on ice and ending up in a serious car accident on a cold winter night. This could be a diabetic losing consciousness due to a severe hypoglycemic episode—I'll call her Sara. Sara crashes into a tree and has severe injuries, including a nearly severed leg, and is thrown from her car into the snow, where she bleeds heavily from her lacerations and quickly becomes hypothermic. Fortunately, when EMS personnel arrive, they are able to quickly obtain her electronic health record on an iPad. The EHR lets them know that she's a diabetic prone to hypoglycemic episodes, so they test her blood sugar and add glucagon to the IV that they're able to set up on the spot. A portable X-ray machine reveals that the most urgent need is to stop the bleeding from her mangled leg, which they quickly tie with a tourniquet. They transfer her to the ambulance, where they can begin to address Sara's hypothermia and start a warm blood transfusion. She has severe facial injuries and will need reconstructive surgery, but the first imperative is to stop her from bleeding to death. Fortunately, the EMS workers are able to e-mail Sara's medical condition and the treatments already begun to the hospital while the ambulance speeds ahead. A few months and several surgeries later, Sara is up and walking around with the help of a motorized prosthetic leg, ankle, and foot. To see her walking down a busy sidewalk, one would never know that only a few months earlier, she had been a hair's breadth from death. Although some of the aftereffects of the accident will follow her throughout life, she is fully aware of the extraordinary treatment she received, and how the omission of any of these technologies could have meant the end of her life. She doesn't pause to reflect on the fact that much of the technology that saved her life was in fact developed by the military.

It may seem counterintuitive that a huge bureaucracy like the Department of Defense would evolve an effective model of rapid innovation, but that's just what the Telemedicine and Advanced Technology Research Center (TATRC) did for several decades following WWII. Karl Friedl, Ph.D., a zoologist by training, research physiologist, and retired colonel in the U.S. Army, led TATRC from its Maryland-based headquarters from 2005 until 2011, during which the center jump-started a dazzling array of research projects. An office of the Army Medical Research and Materiel Command that was created to solve urgent medical problems not addressed by any other branch of government, TATRC specialized in bringing together clinicians, biologists, engineers, mathematicians, and physicists—people who would otherwise barely cross paths—on some of the most innovative research in the world.

I spoke with Dr. Friedl on a cold March day just following the kind of spring snowstorm that makes residents of the Washington, D.C., area groan with weary resignation. He regaled me with almost total recall about the many research projects that got their genesis with the stroke of his pen. Himself an expert across a wide spectrum of disciplines, he has become one of a few people in the country who really understands the organizational dynamics of quickly bringing about radical innovations through the use of converging technologies. I especially wanted to know what TATRC has done right to get such results from an entity as enormous as the DoD.

The first priority, Dr. Friedl told me, is to create a "Senate moment," when the center got members of various specialties who "normally wouldn't talk to each other" together on a phone call or in a meeting. The goal, he said, is to get biologists thinking like engineers, clinicians thinking like physicists, and math modelers thinking like biologists. "It opens up a whole new world for them," he said. "At first they didn't want to go to the meetings, because it was just one more thing to do. But by the end of the first meeting, they were coming up to me and saying, 'When's the next meeting?' It opened their eyes to a whole different way of seeing things." Not only were TATRC's portfolio managers able to ignite interest in a multifaceted approach to solving problems, the organization's decision-making process was kept as streamlined as possible. By comparison, the National Institutes of Health are mired in process, a reality that can, according to Friedl, "kill a good idea a thousand different ways."

Rather than having ideas go through multiple layers of review, as the NIH does, TATRC had only two levels. There was a once-a-week three-hour phone call in which a group of subject matter experts would review proposals and decide whether an idea sounded promising. If it did, it landed on Friedl's desk for approval. There was no need to present preliminary findings or voluminous proposals already critiqued by several committees. If the research project was approved, TATRC would bring together the needed experts, provide a modest, time-limited grant, normally between $50,000 and $300,000 (small by NIH standards), and then apply what Friedl calls "activist management" to keep the project moving quickly.

During his tenure at TATRC, Friedl encountered all kinds of out-of-the-box ideas. "I remember one occasion," he said, "when three guys—physicists from UCLA—called me up and told me they had an idea for X-ray Scotch tape. If they had come in person, I would have laughed them out of my office, because I thought they were joking. 'You're pulling my leg,' I told them, but they said no, they had evidence it would work. And it turns out that when you pull away Scotch tape, in that instant it releases a stream of electrons, which creates a current that actually does X-rays. We gave them seed funding and within a year they had two little rollers that would produce an X-ray. They spun out a company to produce it and asked for more money. At that stage the best thing we could do for them was to help them seek investors."

TATRC often played the role of the angel investor, and when a technology worked, the researchers could run with it like ducklings leaving the nest. Countless innovative projects received their first launch with TATRC funding, but the army has no interest in commercializing the research. "The DoD," said Friedl, "doesn't bring products to market. It fights wars." In 2008, *The New York Times* reported on the X-ray Scotch tape created by the UCLA physicists, wryly calling it "a tour de force of office supply physics,"[3] but it's easy to see how such a technology could become indispensable (no pun intended) in emergency medical situations.

Friedl told me about another project that had met with extreme skepticism in the beginning. It almost sounds like something dreamed up by H. G. Wells, and Friedl said that nearly everyone who heard of it thought it was crazy. "The idea was to take hot air balloons—blimp-like devices—and put communications equipment in them and have them hover over soldiers in battle situations," he explained. "The communications equipment could gather key information from above the battlefield and talk to systems on the ground. Everybody said, 'You're crazy—the enemy will shoot them down.' Of course they could, but the thing is they were cheap and they worked. When they shot one down, you could quickly deploy another one." Friedl called this one an example of the kind of "sporadic innovation" created by a small research project that TATRC was able to launch.

In spite of the long list of innovations that TATRC has spawned that have both military and civilian applications, the agency doesn't consciously seek out civilian crossover potential when considering whether to support research. It focuses, according to Friedl, on military "problems on the ground that need a quick solution." In his writings, he says that every good idea needs a "zealot," someone who is totally obsessed with it, to get it off the ground. This individual, generally a researcher, is typically able to withstand years of harsh criticism when virtually everyone refuses to take it seriously. "The magic number seems to be ten—ten years while everybody shoots it down, and then something happens, the idea takes off, and then suddenly everybody is all over it. Without a zealot to weather all the criticism, the idea will be killed."

Large teams and too-developed processes also kill ideas, Friedl said. "There are many Valleys of Death for an idea. Multiple reviews when too many people can take pot shots can be fatal." He has seen it happen many times. There seems to be a moment in time when the alchemy is just right in terms of an articulated need, a well-designed study, a small team, a limited review process, just the right amount of money, and the all-important tenaciousness of the zealot. When too many people jump on board or the process becomes too complicated, as can be so apparent in NIH-funded basic research, it kills innovation.

TATRC did have one advantage that most research labs could only dream of, and that is the support of Congress. Champions in Congress would shepherd funds to certain projects in the form of earmarks—a process that has met opposition within the current

Congress and resulted in TATRC's downsizing from 180 people to about 60. There is now some uncertainty about how the organization can reinvent itself with a different funding stream and keep its mission of spearheading innovation. The support of Congress can be a fickle, personality-driven thing that easily shifts with the political winds. The congressional earmarks ended in fiscal year 2010, although many projects continue in various forms today.

Even though TATRC never sought out projects with crossover potential, the ability to develop medical innovations that help all of society is what Friedl called simply "the beauty of biomedical research." One surprising initiative that he described to me stood out, and it landed on his desk after the NIH turned it down. It was a project doing research on osteoporosis. The conventional perspective had been that osteoporosis has no relevance for young women when most of them serve in the armed forces. But Friedl knew that about 25 percent of military women developed painful, incapacitating stress fractures during basic training. One of the 120 projects he was managing at the time was on stress fractures, and the two projects just clicked into place. A new alignment was created between the osteoporosis study and the stress fracture study, and Friedl threw all the resources he could behind them. The results yielded some amazing new data on osteoporosis and bone mineral properties. It turned out that (1) young women do indeed have bone issues, (2) taking vitamin D reduced stress fractures by 30 percent, and (3) bone density has very little to do with bone strength. A woman can have very dense but very brittle bones. Perhaps even more surprising was the role of exercise in the building up of bone tissue. It has long been known that women who exercise regularly build up stronger bones, but what the researchers learned is that bone is only rebuilt during the first ten minutes of exercise; after that, the body stops redensifying bone tissue. The conclusion of this work suggests that women who want to build stronger bones should do shorter, but more frequent, bouts of exercise.

I wondered when this information will make it to the lips of doctors who treat civilian women. "It seems to take forever for these things to get into the public perception," Friedl lamented. He wrote a paper about it in 2005, yet my own recent experience shows that doctors are still fixated on bone density tests as the litmus test for osteoporosis. The wide dissemination of new biomedical information is surely a topic for another day, but I was somewhat surprised when Friedl told me that there really is no mechanism for transitioning military medicine to the civilian sector. The process is slow and haphazard.

Most military medical technologies will be quietly integrated into civilian health care, if they haven't already, without any controversy. However, it can never be taken for granted that powerful new technologies will have no downside. The military is also exploring technologies that could easily be abused in both the military and civilian sectors. Drugs are being developed by the Defense Advanced Research Projects Agency (DARPA) that could allow soldiers to stay awake for up to 168 hours without any of the debilitating symptoms of sleep deprivation. This could prove to be a critical advantage in warfare and would dramatically change what the military calls "operational tempo," leading to much longer engagements, where fighting will go on day and night. The research is focused on identifying a "reset button" in the body that is normally activated only by a good night's sleep. It's likely that such drugs will make their way into the civilian market to help the sleep-deprived function for long stretches of time without the need to sleep. The availability of these drugs will almost certainly lead to abuse, and the long-term effects of them are not known. What if the drugs become widely used by college students studying for exams, or by workers trying to gain a competitive advantage over their colleagues? Such use could raise productivity standards to the point where people will feel unable to refuse them, notwithstanding that adequate sleep is indispensable to good health. The result could be a twenty-four-hour-a-day sleepless culture where people are exhausted beyond reason but unaware of it because of the continuous stimulation of drugs.

Another active area of research by DARPA is direct brain-to-machine interfaces, and machines that can be directed by sheer mental power. This technology would involve computerized brain implants that can detect patterns of activity in neurons, translate those patterns into electrical impulses, and send instantaneous commands to machines via wireless transmission. (The same type of technology is now being developed to help amputees and people with motor neuron diseases such as ALS control artificial limbs and machines that can assist them.)[4]

In 2014, Antonio Regalado wrote in *MIT Technology Review* about a DARPA project to develop brain implants that control emotions. The implants—tiny computers—will be able to detect neuronal activity associated with certain feelings, such as fear, and stimulate the brain areas that will block them. This technology, if used wisely and ethically, could advance the treatment of mental disorders such as depression, addiction, and obsessive-compulsive disorder.[5] It might very well be a godsend to people whose symptoms can't be controlled by conventional drugs and talk therapy. However, this technology could also lead to a whole new level of mind control that is frightening to contemplate. What if the implants were used to block fear in soldiers to the extent that they could be sent into situations so dangerous that they would entail certain death? What if they fell into the hands of oppressive governments that would mandate them to make citizens more accepting of their tyrannical ways? And how would they affect people who are already prone to irrational risk taking? What if they were used by criminals to blunt their fear of punishment, making them capable of formerly unimaginable crimes? Or what if they fell into the hands of terrorists, allowing them to not only recruit but manufacture many more suicide bombers than ever before?

Society currently has no adequate safety mechanisms to control such powerful technology because it has never existed before. Mind control, in many different forms, has existed for millennia, but in forms that people could ultimately override. There is no precedent for the physiological control of feelings. In direct brain stimulation, we're dealing with a double-edged sword of remarkable benefits and dreadful risks. Thousands of people with Parkinson's disease or severe depression have now benefitted greatly from deep brain stimulation—the placing of electrodes into certain parts of the brain and sending electrical signals to stimulate those areas. The results have been remarkable for many of these patients. Parkinson's patients have been able to greatly reduce their tremors, and people with treatment-resistant depression have experienced relief that drugs and therapy could not give them. It's because of these benign uses that we shouldn't ban brain stimulation outright, but we urgently need to identify ways that we can prevent its abuse.

In chapter 6, I'll discuss altering the brain in more detail, but here I'd like to emphasize the dangers of such technologies being used by the military. While we depend on our military to keep us safe and are deeply indebted for the services they provide in war zones, we should not forget that war is a devastating phenomenon. The purpose of war is to defeat the enemy through the delivery of the crushing and will-breaking destruction of soldiers, supplies, economies, infrastructures, systems, and governments. It is a terrible fact of modern-day warfare, which confronts the fragmented agents of terrorism, that a great many civilians are maimed and killed in the process. The long-term trend in U.S. military innovation has been to invent ever more focused "surgical" strikes that target combatants and reduce the impact on civilians, yet inevitably, many civilians are destroyed in almost any military initiative. We should not delude ourselves. The radically advanced technologies brought about by converging technologies have the potential for undreamed-of death and destruction when used in warfare and if they fall into the wrong hands.

In 2005, the German physicist and nanotechnology researcher Jürgen Altmann presented a talk at the Nanotechnology in Science, Technology and Society conference in Marburg, Germany. Altmann has been studying the process of military disarmament for thirty years, and he called for the need to limit the uses of CTs and nanotechnology in warfare before there is an all-out arms race for nations to acquire them. Chief among his concerns is that these technologies will inevitably affect society during peacetime, when there is less scrutiny of weapons development, and could actually make war more likely. And even with conventional rules of engagement having been established during wartime, the drive to enact a strategic advantage often leads to nations breaking the rules either secretly or overtly.

Altmann noted that the United States spends about two-thirds of the worldwide investment in military R&D and that about one quarter of federal funding for a National Nanotechnology Initiative goes to the Department of Defense. He reported that "work is being done at universities, the laboratories of the armed services, and the national weapons laboratories on subjects such as carbon nanotubes [superstrong and almost infinitesimally thin structural tubes with advanced electronic, optical,

and thermal abilities that can be used in a huge range of materials],
magnetic nanoparticles, organic light-emitting diodes for displays,
biocomputing [the merging of biology with computer technology],
biomolecular motors [implantable electronic devices so tiny that
they work at the molecular level]. Closer to military needs are
projects on sensors for chemical- or biological-warfare agents or
nanostructured explosives and armour."

Altmann cites the following advances that could easily, and
with great benefit, be transferred to civilian use: Tiny but amazingly powerful computers will be incorporated into weapons,
uniforms, military equipment, and vehicles. Advanced guidance
systems could be included in even small munitions, creating
greater accuracy of strikes. Vehicles could become much lighter
and more agile, and the use of more unmanned vehicles and enhanced communications could save lives. Not only that, nanotech
body implants could greatly increase the safety of soldiers. They
could be used for the monitoring of vital signs, all sorts of communication, enhancement of senses like vision and hearing and
the release of drugs—even those that can cross the blood-brain
barrier. However, once such technologies come into use, they
will almost certainly fall into the hands of enemies and terrorists,
making them harder to defeat and upping the ante in possible
abuses.

In addition, very quick responses programmed into autonomous (unmanned) combat vehicles and systems on the ground, at
sea, and in space, in a highly active battle situation, could lead to
uncontrolled escalation that could amplify battles beyond what
human decision makers would initiate. Altmann is also concerned
that powerful explosives can be made from nanofiber compos-
ites and without any metal components. Such explosives could
elude X-rays and other metal-detecting equipment, allowing
terrorists or criminals to board airplanes and other forms of

transportation, enter buildings, and wreak greater havoc than ever before.

Another fear is that extremely tiny sensors and military spying robots released in the civilian sector could make privacy impossible. As noted above, the novel nature of these technologies means that the military, never mind society in general, currently has no effective process in place to prevent their misuse. Altmann calls for not only international monitoring of these technologies, but the active engagement of the public in a dialogue, and controlled collaboration between the military and civilian sectors.

Such devices as tiny, autonomous sensors the size of insects strike at the heart of democratic values such as privacy and freedom of expression. These devices would allow totalitarian governments and intrusive corporations to monitor a citizen's every move. The technology could lead to great abuse by corporations who gather and use detailed information about us for financial exploitation and for the incarceration of countless political prisoners that a belligerent government wants to control. And criminals would have the means to stalk us, spy on us, and harm us in ways that we currently can't imagine. In his 2005 talk, Altmann recommended the following actions:

- Tiny, self-contained sensors below a few centimeters in size should be banned altogether.
- Metal-free explosives, arms, and ammunition should be banned.
- There should be a ten-year moratorium on body implants and other intrusive technologies that are not exclusively medical in nature.
- Unmanned aircraft and military vehicles and systems should be banned. [This clearly hasn't happened; unmanned drones are being used more than ever.]

- There should be a comprehensive, international ban on space weapons.
- There should be strict compliance and even a strengthening of the rules governing chemical and biological weapons.

I believe that the possibility of international compliance with these proposals is nonexistent, and even Dr. Altmann recognized that it is unlikely. The rapid advance of technological capabilities, the pressures of warfare, the needs of medicine and the free market economy make such containment unrealistic. Although the task seems monumental, our only hope may be to grapple one by one with the potentials of abuse within the realities of the world we live in.

It may even be true that the current generations, the ones making these discoveries, are simply unequipped to formulate solutions to problems they have never encountered. It may require a series of significant disasters before humanity can begin to envision what steps must be taken to ensure that these technologies are a boon and not the bane of human existence. As Albert Einstein said, one cannot solve a problem with the same level of thinking that caused the problem in the first place. It may entail both great triumphs in some areas, such as medicine, and bitter experiences in other areas, such as war, before we learn to control the power unleashed by converging technologies.

The opponents of the radical human enhancement possible using converging technologies are myriad, and they hail from both the left and the right. Antihuman enhancement sentiment is a force in itself. It's not just conservatives like Francis Fukuyama who oppose human enhancement—there is a strong antienhancement movement on the left, which James Hughes, in his thought-provoking 2004 book, *Citizen Cyborg: Why Democratic Societies Must Respond to the Redesigned Human of the Future,* traces back to the horrors of World War II, especially Nazi eugenics and the atom bomb. As these technological horrors unfolded, proscience Enlightenment positivism among liberals gave way to what Hughes describes as "pastoralist visions of a deindustrialized socialism, pseudo-science, spiritualism and back-to-land communalism [which were] all tied up for bohemian radicals with opposition to capitalism."[6] The advance of radical new technologies still arouses suspicion in the minds of many progressives, who associate it with corporate consumerism and paternalism, authoritarian government, environmental destruction, racism, sexism, and all the abuses these phenomena engender. Underlying this distrust, however, is a romanticized notion of the past that stands in contrast to the forces of scientific, corporate, and government oppression.

Ironically, overly romantic notions of the past are also aligned with the sentiments and visceral emotions of Luddites on the right. The fact that extremists on both ends of the ideological spectrum tend to share a sentimentalized vision of the past and deep distrust of current social and economic institutions suggests that attitudes are capable of shifting beyond the well-defined boundaries of left and right. Although each faction is deeply entrenched in its positions on issues like abortion and free-market capitalism, the issues raised by radical human enhancement are exerting new pressures on the human psyche that could lead to new social alignments and configurations. Hughes quotes the New Left activist Todd Gitlin's description of the mind-set of the New Left as "a disbelief in the future . . . We find ourselves incapable of formulating the future."[7]

One of the main issues here is trust. Distrust of humanity is an almost ubiquitous theme underlying the thinking of both bio-

conservatives and New Left liberals. Neither side trusts humanity to make wise choices about the future and neither side believes
in a moral baseline of decency and good intentions in the other.
This creates an impasse that will be difficult to cross.

There is another theme that I return to over and over in regard to technological devices being incorporated into the body.
We already have access to powerful technologies to enhance our
lives, including our knowledge of the world around us, in the
form of things like computers and smartphones. What is fundamentally different about having these technologies incorporated
into our bodies and brains rather than carrying them around in
our pockets? To my thinking it depends on whether, as such
technologies become a part of us, we can override them with our
will whenever we choose or decline to use them at all—in other
words, will we control them or will they control us? And will
they change us in ways that we would not consciously choose
for ourselves? However, we should also recognize that human
enhancement has been going on since the first human stood up
on two legs, from the invention of stone tools, education of the
young, tattoos and cosmetic adornments to selective breeding
(genetic engineering) in the form of arranged marriages and the
selection of mates. It's hard to say that such practices have done
much harm to mankind, and in fact most of them have conferred
great advantages in the struggle for survival.

The libertarian science writer Ronald Bailey responds to C. S.
Lewis's assertion regarding genetic engineering that when one
generation gains the power to change its descendants as it sees
fit, then all successive generations will live in a kind of power-
less servitude to the choices of their ancestors. Bailey responds,
"But when has it not been true that past generations control
the genetic fate of future generations? Our ancestors, too,
through their mating and breeding choices, determined for us

the complement of genes that we all bear today."[8] He rejects what he sees as a kind of fetish for randomness and chance in gene selection, going on to say, "Fortunately, our descendants will have at their disposal ever more powerful technologies and the benefits of our own experiences to guide them in their future reproductive and enhancement decisions. In no sense are they prisoners of our decisions now. . . . Of course, there *is* one scenario in which future generations would be prisoners of our decisions now—namely, if we fearfully elect to deny them access to the benefits of biotechnology and safe genetic engineering."

Some bioethicists and social critics across the spectrum warn of a dystopian future in which society is divided by a kind of genetic and technological caste system. The rich will avail themselves and their children of every possible enhancement, making them healthier, smarter, stronger, and longer-lived, while the poor languish in a kind of biological quagmire of inferiority. The result will be two sharply divided classes of humans. The biologically and technologically enhanced will regard the "naturals" as mere savages fit only for the grunt work needed to support them and may even find them so expendable as to commit wholesale genocide against them. Or, the "naturals" just might get wind of their impending fate and commit preemptive genocide against the enhanced. Either way, we can expect a worldwide catastrophe. But this view doesn't take into account the fact that, over time, all new technologies started out being available to the rich and then filtered down to the not so rich, nor does it take into account the values of advancing democracy, which seeks to provide mechanisms for providing equal access to all technologies. Government programs like Medicare and Medicaid in the United States and the National Health Service in Britain could greatly enhance the equal distribution of medical treatments like brain implants for people with dementia or mild cognitive disorder. An important question that we will need to face earlier rather than later, however, is whether government and private health systems (including private insurance companies) will cover what they regard as high-technology "enhancements" without an underlying "disease" to treat.

In the United States, we're already in a gray area when it comes to access to drugs and procedures that provide a form of physical or mental enhancement. The deciding factor seems to be whether we have what social and medical consensus accepts as a disease that needs to be treated. Excessive shyness, now called social anxiety disorder, can be treated with Paxil to help sufferers become more outgoing and socially at ease. Although some people oppose such treatment on the grounds that shyness is simply a personality type, not a disease, the millions of people who have sought relief from the condition in drugs like Paxil are saying that whatever you call it, shyness is a painful condition that they want to be free of. At any rate, the government and insurance companies seem to concur that social anxiety is indeed a disorder because they cover drugs to treat it. Drugs like Viagra fall into the same category. It used to be accepted that many men lose sexual potency as they grow older, but now we have a condition called erectile dysfunction that can be treated with drugs. Meanwhile, insurance companies that cover reconstructive plastic surgery after an accident do not cover face-lifts. Having an aging face is not regarded as a disease to be treated, but will this always be the case as society shifts its criteria to whether or not a condition causes psychological suffering? Recent history has shown that over time more and more conditions have quietly been moved into the "disease" or "disorder" category and thereby lent themselves to correction. Using Viagra as an example, might not the entire spectrum of the signs of aging become amenable to treatment, including those that cause social and emotional distress?

Much of the course of whether or not, and to what extent, we come to accept treatments directly related to aging depends on our social attitudes. Is aging a disease? Biology and the entire field of gerontology certainly approach it from a medicalized standpoint. And American society is famously youth obsessed, as is obvious in our popular culture. In chapter 7, I'll explore this issue in more detail, but here I use it to illustrate a pivotal point.

For something to be considered a disease, it must stand in comparison to some version of what is "normal." When we allow or even promote treatment of conditions like shyness and melancholy, we assent to the view that to be normal, one must be reasonably outgoing and upbeat. At least, this is what our culture seems to have accepted to the point that it goes without saying. In the determination of whether to treat them, the key issue is whether the conditions of shyness or melancholy are *experienced as unwanted suffering* by the person who has them. This is the litmus test broadly accepted by the American Psychiatric Association's (APA's) *Diagnostic and Statistical Manual of Mental Disorders* (*DSM*), the guidebook that sets the standard for treatment, including insurance coverage, for conditions deemed to be mental disorders.

While not laying out any specific definition of "normal," the manual recognizes that mental health is not a clearly demarcated phenomenon, that it lies upon a wide spectrum, one end of which we consider healthy and another end that entails severe pathology. The latest edition of the book, the *DSM-5*, released in 2012, recognizes for the first time many different shades of severity in mental disorders. In the *DSM-5*, pedophilia, for example, is not classified as a disease in itself, but the unwanted feelings of distress caused by feelings of attraction to minors is classified as an illness—Pedophilic Disorder.[9] This particular decision on the part of the book's editors was the subject of almost hysterical opposition until the APA made it clear that it unambiguously considers acting on pedophilia-related feelings a crime. Such debates in the field of psychiatry are by no means over and show every sign of continuing into the foreseeable future.

While society's institutions such as the APA have a powerful effect on what conditions will be treated, one can never underestimate the demands of the marketplace in a capitalist society that can institute changes of its own. Cosmetic surgery is a powerful example. In 2012, Americans spent $11 billion on 1.3 million cosmetic surgery procedures such as breast augmentations, liposuction, face-lifts, rhinoplasty (nose jobs), and injections of Botox and fillers.[10] Given the sharp rise in cosmetic procedures, which are not covered by insurance, in recent years, one can only assume that there would be even more procedures if more people could afford to pay for them out of pocket.

There are strong social pressures, such as the demands of patients, to provide more treatments for more diseases, but there are also strong social pressures that decry any kind of enhancement, especially in the sphere of mental enhancement, whether by drugs or implants. Aesthetic enhancement, while widely accepted in the form of moderate enhancements such as hair coloring and cosmetics, nevertheless meets with mixed reactions when it advances to things like face-lifts and cheek implants. It's a fine line that women, in particular, have to walk these days between being pretty enough for social acceptance and going "too far" with things like liposuction, lip jobs, and face-lifts. If this were not the case, the increasing number of women and men who receive cosmetic surgery procedures wouldn't feel compelled to keep their procedures secret. We live in a society that can't seem to decide what's more important—fitting a certain aesthetic standard, or being "natural." The question of "authenticity" is one that we give great lip service to, but to which we are far less devoted in practice. The most contentious debate rages over augmentation of the most important organ—the brain. There is no disagreement that the brain is the seat of our identity, and any meddling with it is potentially dangerous and capable of creating great inequality.

The German conservative philosopher Bruno Macaes writes, "If someone were by means of a brain implant to acquire superhuman memory, it would be foolish to be impressed by his displays."[11] Perhaps, but it's difficult to tease out why using an implantable technology to improve mental functioning and attain extra knowledge is different, in effect, from achieving the same ends through memory training and education. Macaes and others like him might argue that there is something valuable in the sustained and dedicated effort to learn new things by conventional education methods with the brain one was born with. I can certainly see his point. The experience of striving so often results in benefits in addition to those we were specifically striving for—patience, perseverance, the ability to pursue delayed gratification, self-denial and the real enhancement of one's self-esteem, pride in accomplishment, and self-confidence. But there's no reason to believe that a person couldn't still attain the same benefits when striving from a higher platform of ability. There's nothing inherent in technological enhancement that rules out the human struggle; it just offers the possibility of transferring that struggle to a higher and more efficient level.

Macaes further worries that even the scientists working on human enhancement innovations lack "a genuine understanding of the most fundamental processes," and that no one "is in a position to look at biotechnological enhancement from the outside and judge its correctness: no one is in charge."[12] Again, this may very well be true, but the same trepidation could have been, and no doubt was, applied to the first surgery, the first vaccination, the first trip into space, and the first exploration of the New World. No one really knew how these enterprises would pan out until a leap of faith overcame the fear of doing them. Large and ambitious innovations all, by definition, take us into places we have never been before. Should we foreclose any bold new enterprise because we are unwilling to take any risks?

Macaes reports on Ronald Dworkin's observation in the book *Life's Dominion,* where he asserts the value of "the idea of the sacred in terms of a process or enterprise requiring long labor and great effort, the sort of investment that is impossible to replicate in an age of technical reproduction."[13] But one can look at the age of calculators and computers, when human productivity was dramatically increased over that of previous eras, and see that our lives have not become easy or free of struggle. Rather, the context of struggle has changed, shifted to a new plateau where more can be done in a smaller amount of time, but along with this development, there is more pressure to accomplish even more in even shorter spans of time. Fewer people till the soil, but no one would regard the average office job as free from effort and stress. It's possible, of course, that should progress proceed at the same, or, as Ray Kurzweil predicts, an accelerated pace, until most of the world's material needs are met, people won't have to struggle as hard for survival or their material needs. However, there is still an unimaginably long future in which humanity might strive for social, intellectual, cultural, artistic, and spiritual goals.

If the past is any indication, each new leap in ability will open up new challenges and new avenues for achievement. The human condition, in its essence, may not change appreciably at any time in the foreseeable future. Should there ever be an appreciable change in the human condition, it is likely to be met by human beings, or transhuman beings, very different and better equipped than us to deal with it because they will have traveled so much further into the human adventure.

Authenticity is a value that Macaes feels would be lost for those who embrace radical enhancement, which he sees as cheating in the game of life while simultaneously losing something precious. He writes,

> Consider someone who is happy and feels pleasure because things are going his way. He has made the world an extension of himself. Nothing happens which he could really wish to be different and he knows this because he made it happen himself. He has learned the art of mastering human situations. Now compare him to the poor fellow who feels just as happy because he has taken a powerful mood brightener. Whatever happens to him is not a result of his actions. Events would make him very unhappy were it not for the fact that he no longer cares for events. Events are now so absolutely irrelevant that he will be happy no matter what happens. They are part of the external world and operate according to forces that he does not master because he never learned how to master them. . . . Most of what we have come to regard as the promise of biotechnological enhancement is a radical betrayal of the modern technological project. Its means and its ends turn us away from the world and the drive to master it.[14]

Macaes makes some strong assumptions. One is that the person who mastered life so well has not been naturally endowed with high intelligence, a sociable personality, great emotional resilience, and a positive formative environment, whereas the person on mood brighteners was not so endowed or blessed and may actually have been depressed. He assumes a completely level playing field in which the first individual simply exercised wisdom and integrity while the second fellow was too lazy to master life and instead decided to cheat. Regardless of our moral judgments, doctors don't currently put healthy, asymptomatic individuals on "mood brighteners"; antidepressants are given to those who suffer symptoms of anxiety and depression. Prescribing medications for the depressed person is really just intended to help level the playing field. And taking medications that improve his coping mechanisms wouldn't wipe away an individual's need to master the situations of his life—they would just make him as capable, or more nearly so, as the naturally well disposed.

The second assumption is that the person who by all outward appearances has mastered life has no problems, no private anguish, and no worries to contend with. We should all know to resist such conclusions by now—witness famous celebrity suicides, such as the 2014 suicide of the wealthy, successful comedian and actor Robin Williams, a beloved man who by society's standards had reached the very apex of success in life. It's now known that Williams had developed a devastating form of dementia known as Lewy body disease, a condition that his wife believes caused him to commit suicide.

We might liken our conception of what life may be like for the transhumanists that come after us to the way we viewed adults when we were children. We watched our parents come and go as they pleased, stay up as late as they wanted, spend money as they deemed appropriate, drive cars, and enjoy all sorts of freedoms that we, as children, were denied. We saw their freedom and power, but not their cares and worries, the heavy responsibilities that go hand in hand with adulthood. It would be just as naïve to assume that a person with higher intelligence, greater physical vitality, and a much longer life would have no real problems or challenges. At the same time, not having to expend the same efforts on combating the hardships we now face, such as debilitating diseases and the ravages of aging, could well liberate us to work on higher issues and problems that are more social, intellectual, and spiritual in nature.

The question of human enhancement today seems to pervade many people's most catastrophic doomsday predictions while infusing others with great optimism and hope for a brighter future. The differences in opinions about the desirability of radical human enhancement might be as intractable as the differences between optimists and pessimists. Optimists, it has been amply demonstrated, have a bias in favor of anticipating positive outcomes, while pessimists have a bias in favor of believing in negative outcomes. Psychology has proven that our mind-set, whether positive or negative, tends to actually change outcomes to make them live up (or down) to our expectations—in the form of self-fulfilling prophecies.

Ronald Bailey asserts that "the inviolable core of our identities is the narrative of our lives—the sum of our experiences, enhanced or not." He goes on to say, "But whoever we are persists and perhaps even flourishes if we choose to use biotech to brighten our moods, improve our personalities, boost our intelligence, sleep less, live longer and healthier lives, change our gender, or even change species."[15] In other words, by and through our choosing, we become more, not less, ourselves.

I suspect that we are more than the sum of our parts, even our heredity, life experiences, and accomplishments so far. It may be in the nucleus of our desires that we find out who we truly are. Life is, in the last analysis, a perpetual state of becoming—becoming more, becoming better or worse, often becoming someone who is in many ways different from the person we started out as or who we thought we were. Who doesn't wish ardently for some qualities and abilities she wasn't born with? Don't our actual choices mean as much about us as the attributes randomly bestowed upon us? As individuals and as a species, there may be no other way of knowing than to take the journey of human enhancement to see who, and what, we can become.

# 6

## Building a Better Brain

Neurologist Douglas Scharre's practice is largely devoted to treating Alzheimer's patients, but if he never had to treat another one of them, he would be "as happy as a clam." Watching the progression of this devastating condition in patient after patient and having only limited tools with which to treat them is discouraging. But the down-to-earth medical director of Ohio State University's Memory Disorders Clinic with the gentle, self-deprecating laugh is still hopeful that better treatments can be found. It's for this reason that he has teamed up with Ohio State's neurosurgeon Ali Rezai to conduct a small study using a form of brain "pacemaker" to treat patients with early-stage Alzheimer's.

There is no more dreaded malady than Alzheimer's disease (AD). Americans fear it more than cancer, heart disease, or any other condition you can name. Already more than four million Americans live in the mind- and memory-stealing fog that will slowly progress to complete dependence, and, eventually, death. That's enough people to populate a major city, and with the aging of the baby boomers, AD threatens to overwhelm our health care system with nine million patients by the year 2040.[1] The plight of family caregivers is almost as dire as that of the patients, hence Nancy L. Mace's bestselling book *The 36-Hour Day*, which awakened the world to the all-consuming job of caring for a loved one with AD.

It's thought that the physiological brain changes associated with Alzheimer's begin many years before symptoms are noticed. Beta-amyloid, a naturally occurring protein in the brain, begins to fold improperly and then to clump together in sticky plaques in the spaces between brain cells. The cells themselves become clogged with neurofibrillary tangles, caused by another protein, called tau. It isn't known whether this buildup of protein fragments causes the symptoms of Alzheimer's, but soon brain cells start to die, most markedly in the hippocampus, a structure that is vital to encoding memories. Over time, the cell death spreads throughout the brain. Brain matter shrinks as more and more cells die and the memories, ability to reason, and personality dim until even brain stem functions like swallowing are lost. It can be a long, agonizing process that is inevitably fatal.

One of the most terrorizing parts of Alzheimer's is that it begins with mild forgetfulness that can be hard to distinguish from normal, age-related changes. Almost everyone over fifty, when they misplace their car keys, will experience a moment of terror when they wonder if they have the beginnings of AD. In most cases, it's not Alzheimer's. Doctors make distinctions such as: If you forget where you left your car keys, it's most likely not AD; if you forget what the keys are *for*, it could be. Over time, memory problems get worse and symptoms such as agitation, delusions, and hallucinations creep in. There are medications that can slow deterioration for a period of time, but nothing stops the steady downhill march of the disease. In the advanced stage, the patient suffers less than his caregiver, not recognizing anyone, not understanding what is happening to him, and enveloped in a miasma of forgetfulness. The oldest memories are the last to go, and often he truly does experience a "second childhood," being immersed in his early memories, which are still intact. This stage is particularly painful for caregivers, who have long ago lost a person while the body lives on. Death is often a relief, but caregivers can be left physically, emotionally, and financially drained, with a complicated grieving process to boot.

In a telephone interview, Scharre was careful not to use the word "cure" as the stated goal of the pacemaker study, which employs deep brain stimulation (DBS) applied to the brain's frontal lobes in the hope of preserving intellectual functioning longer in AD patients. Because of the overwhelming death of brain cells at the late stage of the disease, no treatment undertaken then could restore brain function. But if caught early, the available treatments work better and can stave off worsening disability for a matter of a few years—no small gift to patients and their loved ones. "If you get it at eighty," he told me, "you may not make it to eighty-five. This means that if you can stave it off for five years, you've cut it off." The patient might die from another cause before the AD destroys her personality, cognition, memory, and ability to function. Her last years might be far happier and more meaningful for her and her caregivers.

Three drugs aimed at improving memory—Aricept, Exelon, and Razadyne—have been FDA-approved and can make a measurable difference if started early enough in the disease process. Less well known are the psychotropic drugs such as mood stabilizers, antidepressants, and antipsychotics that treat the behavioral disturbances that come with the middle stage of the disease. Scharre identified them as "anger, angst, and agitation," and wondered why so many people are opposed to the use of drugs to treat them. "The very word 'psychotropic' is controversial to some people," he said, "but all it says is that the drug works on the brain. In the case of a brain disease, you *want* drugs that work on the brain. If you have heart disease, you wouldn't hesitate to take medication for it. Why would the brain be any different?" I suggested that the highly vocal antipsychiatry movement, which decries the use of any drugs affecting the brain, might have something to do with it. "Some very smart people have these attitudes," he said, "but the drugs help. They should be embraced and they should be started early. We hope to find more, and more effective, drugs."

The other arm of treatment for AD includes behavioral therapies like physical and mental exercise. "I encourage people to do things that activate the brain," he continued. "Get out and do something. It doesn't matter what, as long as you're doing something." Challenging oneself through activities, it has been shown, promotes neurogenesis, or the birth of new brain cells and the sprouting of new connections. Practicing the piano and playing sports nudge our brains into neurogenesis mode and so, it has been suggested, does DBS.[2] In fact, promoting neurogenesis through any and all means could help the brain build up "reserves" that could postpone the progression of functional disability in AD.

Electrical stimulation of the brain has been used to treat various diseases, including psychiatric conditions, for over two hundred years. It wasn't until the 1960s that scientists established its benefits in neurodegenerative diseases like Parkinson's disease, essential tremor, epilepsy, and even obsessive-compulsive disorder. To date, about 120,000 people worldwide have received DBS, and the procedure, while invasive, has been shown to be quite safe. There is about a 1 percent incidence of stroke or cerebral bleeding during surgery to implant electrodes in the brain and a 5–10 percent risk of complications due to problems with the hardware such as breakage or infection.[3] However, if something goes wrong once the electrodes have been implanted, the mild electrical current being delivered can be turned off without surgery.

DBS, which delivers continuous electrical impulses to the desired region or "node" in the brain, affects the circuitry, or patterns of communication pulsing through nerve cells and synapses. It can either ramp up the flow of signals passing through circuits or slow them down, depending on whether the region is overactive, as in depression (more later about that), or underactive, as is the case with a brain region in Parkinson's. The view of Alzheimer's as a "circuitry" problem is somewhat novel but is gaining attention in the research community. After all, the brain is anything but a static organ, and consciousness is characterized by the constant flash of electrochemical signals through the brain's many neural pathways. Multiple "circuits" are formed through the latticework of connections among brain cells. It has been theorized that a change in one part of a circuit may affect the behavior of many other nodes and circuits downstream of the change, so intervening at just the right spot could have both risks and advantages.

I asked Scharre why his team decided to focus on the frontal lobes as a possible site for intervention in Alzheimer's disease. It turns out that in each frontal lobe there is a node, or major "switching station" of currents from various parts of the brain, that seems to play a prominent role in alertness, attention, focus, and decision making—the so-called "executive function" of the brain. The hypothesis is that stimulating this region could lead to better functioning overall. It may seem more intuitive to focus on the hippocampus, a structure that encodes incoming information so that it can be stored in long-term memory. However, the cells of the hippocampus start to die very early in AD and form what Dr. Scharre calls a "dying zone." "In AD," he said, "the pathology starts in memory circuits and spreads predictably to known neural pathways. I thought, 'let's bypass the dying zone and go to the next node, the one getting signals from the frontal lobes.' "

The science behind brain implants is moving fast. Dr. Rezai, who has implanted electrodes into three AD patients so far, is already looking ahead to the next generation of brain implant. The current model includes a small pulse generator, which is implanted just below the collar bone and connected to thin wires that run under the skin, up the neck, behind the ears, and up through the top of the head, where two holes are drilled into the skull. The wires are then inserted deep into the brain, to a position just behind each side of the forehead, while the patient is fully awake (the brain feels no pain, just the pressure of the drilling). Then, mild electrical pulses stimulate the targeted regions.

This model does not collect information from the brain, as future models will, and after three or four years the battery in the generator will give out. When this happens, the generator is replaced—a minor surgical procedure that can usually be done on an outpatient basis.

The next brain pacemakers will actively monitor brain signals and, when needed, respond with the correct amount of stimulation to normalize brain activity. This is already being done in patients with epilepsy, and it's proving effective at preventing seizures. While this technology represents a quantum leap in the ability to manage diseases of the brain, it also gives pause to some bioethicists, who worry about the collecting, storing, and possible misusing of the most intimate data about us of all—our brain activity.

Drs. Scharre and Rezai have yet to publish their findings as of this writing, but they may have done so by the time this book comes out. No one at this point is claiming that DBS, or any other available treatment, has the potential to prevent or cure

Alzheimer's, and with lengthening life spans, AD threatens to become a global crisis. It's predicted that worldwide, in the next forty years, there will be 115 million new cases of the disease. That's comparable to the population of an entire nation, and it's a burden that no one yet knows how to pay for. If radical life extension is just around the corner, it may be meaningless or even disastrous without a way to eliminate AD. Current treatments are still highly meaningful to elderly patients who may not live long enough to experience the worst of AD. But if one develops Alzheimer's at 75 and lives to be 130, 140, or older, what then?

Another issue that bioethicists are swift to point out is the possibility that, should DBS prove to markedly improve brain function, healthy people will start to seek it out. How will we draw the line between those who are eligible due to some level of cognitive impairment and those who suffer from simple age-related memory problems? Both Dr. Scharre and Dr. Rezai emphatically stressed that the DBS they are using in Alzheimer's patients should never be used in a healthy person. But it's not hard to imagine that people who just want to boost their brain-power will seek it out. The question is how to regulate such technologies within an economic free market and societies that are embracing ever-greater degrees of personal choice when it comes to enhancement.

*Forbes* reported in a recent article that Americans already spend $1.3 billion a year on brain-training products, and this doesn't even include the $1 billion spent on nutritional supplements promising to boost or protect brain function and an unknown amount on cognition-boosting drugs like Ritalin and Adderall.[4] Brain enhancement is big business, and as the baby boomers age, it is expected to get much bigger.

James Hughes provides an overview of the array of cognition-enhancing technologies already available to consumers on the Institute for Ethics and Emerging Technologies Web site. One is a neurofeedback device that can be worn for hours on a daily basis to help train people to recognize and maintain mental states such as focus, memory, and attention. This technology is affordable, but it's hard to say whether and how much the technology is helping, and it's unlikely that anything deemed by its manufacturers as safe to be worn for several hours a day is doing anything at all. Another legal and available technology is called transcranial direct current stimulation (tDCS), which trains low-level electrical currents directly on the scalp and claims to improve math, language, and learning skills. Research on this technology is mixed, but critics note that the few studies done on it used not only tDCS but rigorous brain-training exercises concurrently, and it's hard to say whether improvements in cognition are due to the rather silly-looking headsets or the exercises. The headset sells for about $250, and it is unknown how prolonged use of it might affect the brain.[5]

An FDA-approved technology that is available for the treatment of depression is transcranial magnetic stimulation (TMS), which was invented in 1985. It may sound like something out of nineteenth-century quack medicine, but it has been shown to not only lift depression but also to improve memory and speed of recall. TMS works by placing large electromagnetic coils (or magnets) on the scalp and is thought to both depolarize neurons and to stimulate an increased release of nerve growth factors that can lead to neurogenesis. DBS, already being used for Parkinson's disease, obsessive-compulsive disorder, and tremors, is being tested not only on AD patients but on those with Tourette's syndrome, chronic pain, and treatment-resistant depression. Now the U.S. military is funding some of the most cutting-edge brain research. Because of the high incidence of military veterans who have served in Iraq and Afghanistan suffering traumatic brain injuries (TBIs), DARPA has thrown its heft behind the race to develop permanent memory-enhancing brain implants.

Since 2000, 280,000 military members and 1.7 million American civilians have been diagnosed with TBI. Typically, the condition makes it very hard to access memories formed before the injury, and equally difficult to form new memories after the injury has taken place. Memory deficits like those seen in TBI can wreak havoc in a person's life, disrupting even simple memories like how to dress or tie a shoelace. It's usually a long, slow road to recovery using physical, mental, and occupational therapy, and recovery is often partial. If the DARPA project (called the Restoring Active Memory or RAM project) pans out, TBI patients will soon have a permanent, self-enclosed brain implant that reads and interprets brain signals, then reacts with lightning speed to stimulate neurons that will access the needed memory. The implant is intended to act directly on the hippocampus and an area called the entorhinal cortex, and to aid in both creating new memories and accessing older ones.

The RAM project is a tall order, one requiring the collaboration of brain researchers, medical doctors, mathematicians, physicists, psychologists, and bioengineers as well as computing specialists, and it's intended to take the brain implant to the next level. The implant will consist of a tiny pair of high-density electrode arrays made of flexible, biocompatible polymer that will provide both the recording of brain signals and neural stimulation from sixty-four channels. From deep inside the brain, the implant will communicate wirelessly with a small electronic component worn around the ear that stores digital information associated with memory storage and retrieval, moving the technology well into the terrain of direct brain-to-machine interface. The RAM device isn't being designed to help recipients recall all of the words of Shakespeare's *Macbeth,* for example, but to help them recover "task-based motor skills" related to "life and livelihood." In other words, it's hoped that the device will heal soldiers well enough to put them back into the field, flying airplanes, operating equipment, and making combat-related decisions, as well as to go home and resume their civilian lives.

The entorhinal cortex, which is the main focus of stimulation in the RAM project, is described by lead researcher Dr. Itzhak Fried as the "golden gate to the brain's memory mainframe," playing an integral role in capturing and encoding experience into lasting memories. Dr. Fried, a professor of psychiatry and biobehavioral sciences at the University of California, Los Angeles, has teamed up with UCLA's Mayank Mehta, a professor of physics and neurobiology, and Harvard neuroscientist Gabriel Kreiman to record the activities of small neuronal populations and even single neurons in epilepsy patients that already have electrodes implanted in their brains. Under the DARPA grant, they will next develop computational models of the various activities. These models will help the devices recognize certain signals in the brain and then amplify them, thereby "prioritizing" them for encoding into the long-term memory. The sixty-four-channel device, only about one millimeter in diameter, will be created by bioengineers at the Lawrence Livermore National Laboratory in California under a separate DARPA grant.

This enormous project, which has brought together such a diverse array of specialists, is yet another example of the U.S. military's ability to bring about major technological advances in short amounts of time. DARPA, which awarded the sizeable grants in 2014, has given researchers four years to develop the technology and create an implantable device for humans. Considering that the science has already provided proof of principle in rats and monkeys, the chances are good that a next-generation memory-enhancing brain prosthesis will be available by 2018. This leads to the question, What is the future of memory-enhancing brain implants? Will these implants become available to people with normal memory who want to learn a new skill or language, enhance their academic test scores, or simply gain a competitive edge in the workplace? Although the doctors I have spoken with have stressed that so far the technology is only being considered as a medical therapy, a great many questions remain open.

In fact, very little is known about the above-mentioned technologies and how they might be working. No one seems to have a detailed understanding of why transcranial magnetic stimulation may help to lift depression, or what the long-term effects might be of stimulating certain nodes in the brain and not stimulating others. If brain implants only help to store and retrieve useful memories, their effect will be highly benign; but what if they end up amplifying unwanted memories in people with PTSD, anxiety, or other mental disorders? The act of forgetting, as deeply entwined as it is with our mental health, may exist for a purpose. And as for short-term memory, there may be very good reasons why we can only hold a limited amount of information in our consciousness at any given time. Witness cases of autism in which the individual has a photographic memory that recalls all kinds of facts in minute detail, but is lacking in critical skills like decision making and daily self-care. If the mind were to become flooded with too many conscious memories, would decision making and other executive functions be impaired? Too little is known about brain function at this time to easily predict what radical memory enhancement might do for us.

The current focus on ameliorating disease and disability with brain implants suggests that society is not rushing headlong into the enhancement of healthy brains. But given the exponential nature of converging technologies to bring about dramatic leaps in technology, the next generation of technologies is likely to catapult us into bioethical territory that only a few people have even begun to think about. Critical to the process of innovation will be control of the private sector, in which the drive for profits within a short-term time window is exerting great pressure on companies to gain the intellectual property, patents, and market access to commercialize brain-enhancement technologies. A system of regulations needs to be put into place to manage the long-term effects of technologies that have barely begun to take shape. It's as though we're already driving on a road that's being built one brick at a time while we drive on it, and this phenomenon will only become more acute as CTs build on each other, one quantum leap leading to another as we embark on a future that no one can adequately anticipate.

One of the most significant challenges in regulating the implants and technologies based on CTs is that many of them defy neat classification into categories like "treatment" or "enhancement." In many cases, the technologies that treat also enhance, with no clear dividing line to alert us as to when one has crossed over into the other. It's entirely possible that the widespread use of memory implants is inevitable, first in older people, and this will lead to a "new normal" in what kind of memory should be expected in an eighty-year-old. Memory prostheses may even become as common as glasses and contact lenses. In the past, no one expected anyone over fifty to have twenty-twenty vision; today less than twenty-twenty vision is a seen as a medical condition, an impairment that must be treated in order to drive a car. If we are to extend our legal and regulatory framework to include human enhancements, we must either "medicalize" conditions previously considered normal or place enhancement, no matter how inadequately defined, into an entirely new category with new standards and guidelines.

Another issue that at present shows no sign of being resolved is the ability of individuals and governments to hack into any computerized device or component. It may become possible in the near future for nefarious individuals or agents to disrupt medical implants or to steal information from them. So far, computer hackers have proven themselves capable of accessing and stealing all sorts of encrypted information, causing financial, privacy, and reputational harm to people and organizations; it's only a matter of time before actual people can be harmed through computerized components that their bodies depend upon. If neural implants can obtain desired effects by stimulating certain areas of the brain, just imagine the havoc that could be wreaked if criminals or a belligerent government could attack or manipulate people's brains in real time.

Then there is the familiar problem of vast amounts of almost unimaginably intimate data—obtained from the inner recesses of our brains—that could exist and be stored somewhere. Just as the providers of our cell phone services now have records of almost our every move, the providers of brain implants and the data flowing through them would have records of our very thoughts. Would they then "own" this data? Could such information be bought, sold, or shared with law enforcement or the government? What legal or governmental bodies would regulate its uses, and with what powers of enforcement? How would we even know if a transcript of our mental activities was being accessed and perhaps manipulated? Would the laws to regulate this data be applied internationally or would they vary from country to country? Some bioconservatives use such questions as reason enough to place a moratorium on the creation of brain implants. They typically envision a "Big Brother" type of government that enforces conformity of behavior, while I envision our brains being used as commercial billboards by companies who want to sell us everything from soup to nuts. Neither is a pleasant scenario.

In 2003, Ronald Bailey examined the most common objections to cognitive enhancement for an article posted on the Center for Cognitive Liberty & Ethics Web site that is still relevant today.[6] Bailey deconstructs eight bioconservative arguments and finds none of them especially compelling. At the time his article was written, the most widespread cognitive enhancers were drugs like Prozac and Ritalin, but the arguments remain the same even after over a very active decade of research into all kinds of cognitive enhancement technologies. The first argument is that neurological enhancements must be avoided because they permanently change the brain. Bailey points out that traditional activities like teaching also permanently change the brain, as does practicing a new skill or learning a new language. If Ritalin makes a student more studious and attentive, the acquisition of knowledge enabled in him is just as valid as that gained by a child who is naturally studious and attentive. If permanent change to the brain is what we want to avoid, we will have to forego math classes, tennis lessons, and crossword puzzles as well as drugs and brain implants.

The second argument Bailey cites is that neural enhancements will lead to an unegalitarian world where intellectual castes align with economic advantages. This is one of the first objections to cognitive enhancement in any discussion and it is widespread among both liberals and conservatives. The fear is that not only will the rich and their children be much smarter and more competitive than the poor, rich nations will exercise gross economic advantages over poor ones. But as Bailey and others have pointed out, cognitive enhancements are actually more likely to create a *more* egalitarian society in the near term. The reason is that enhancements are being developed and distributed first as therapies. The benefits of enhancement, especially if it becomes a prescription covered by medical insurance, will likely equalize intelligence levels by bringing those with impairments into the "normal" or "average" range. This means that more naturally endowed individuals (or those with the educational advantages that money can buy) will have less of a relative advantage over others, not more. The advantages of having smarter citizens will benefit society at large. If truly universal health care lies in our future and cognitive enhancement becomes a covered health benefit, no group will have a cognitive advantage over any other group. But it is likely that there will soon be a new normal that people must live up to.

The third objection to cognitive enhancement is that such technologies will be self-defeating to those who choose them because society will become *too* egalitarian. When enhancement is widespread, no one will have a competitive advantage, and this will render the technology worthless. This rather cynical view assumes that our only reason for enhancing our brains is to lord them over others; it overlooks a whole spate of reasons, not the least of which will be the enrichment of our lives on every level, for which we would choose to enhance our brains. Bailey notes that even if no one's social position is changed by enhancement, "the overall productivity and wealth of society would increase considerably, making everyone better off. Surely that is a social good."[7]

The fourth observation is that cognitive enhancements will be difficult to refuse because of social pressure to remain competitive in a world of enhancers. Bailey writes that "it's not as if we don't all face competitive pressures anyway—to get into and graduate from good universities, to constantly upgrade skills, to buy better computers and more productive software. . . . Some people choose to enhance themselves by getting a Ph.D. in English. . . . It's not clear why a pill should be more irresistible than higher education, or why one should raise special ethical concerns while the other does not." He quotes psychologist Michael Gazzaniga, who points out that simply acquiring extra knowledge is no guarantee that one will excel in anything. Gazzaniga says, "I know a lot of smart people who don't amount to a row of beans. . . . So a pill that pumps up your intellectual processing power won't necessarily give you the drive and ambition to use it."[8]

But neurological enhancements will undermine good character, argue bioconservatives in the fifth objection. Fukuyama asserts, "The normal, and morally acceptable, way of overcoming low self-esteem was to struggle with oneself and with others, to work hard, to endure painful sacrifices, and finally to rise and be seen as having done so."[9] Bailey replies that there will be plenty of challenges for the cognitively enhanced to prove their mettle. "Cars, computers and washing machines have tremendously enhanced our ability to deal with formerly formidable tasks," he says. "That doesn't mean life's struggles have disappeared—just that we can tackle the next ones."[10]

The sixth objection is that neurological enhancements undermine personal responsibility, but Bailey points out that availing oneself of enhancement in some area in which one is lacking is actually the responsible course of action. The changed behavior of a child with attention-deficit/hyperactivity disorder (ADHD) who takes Ritalin is what's important, not how easily the child was able to change with the help of a drug. Perhaps having the humility to admit that one needs a drug is the moral litmus test. It's the decision to take Ritalin that shows moral responsibility, and such decisions should not be discounted.

Next there is the objection that cognitive enhancements enforce "dubious norms." Here we have an objection to what will increasingly be higher norms of functioning, an argument that I find confounding. After all, does anyone object to the norm of correcting one's vision to twenty-twenty, even when so many people need corrective lenses or laser surgery to achieve it? Would society be better off if we held the norm to the vision of the average forty-five-year-old in 1787? Bailey is highly suspicious of barriers that would prevent people from aspiring to a higher norm than what nature has bestowed upon them. He writes, "Change may come, but real people should not be sacrificed to some restrictive bioethical utopia in the meantime. Similarly, we should no doubt value depressed people or people with bad memories just as highly as we do happy geniuses, but until that glad day comes people should be allowed to take advantage of technologies that improve their lives in the society in which they actually live."[11] It's interesting to note that there are so many cries to abolish the pressures to enhance oneself by people who nevertheless impose substantial social pressure on others not to enhance themselves. Either way we are subjected to pressures to conform to society's norms, not our own.

The last objection is more or less ubiquitous throughout any discussion about enhancement of any kind. It is that neurological enhancements will make us inauthentic. Bailey asserts that just the opposite is true—our choices are a more reliable expression of who we really are than the lottery of genetics or circumstances. Furthermore, underlying this objection is the assumption of both a static, unchanging self and a bona fide "natural" state. There is currently no agreement on either the concept of an unchanging self or the idea of an entirely "natural" self, since education, culture, history, and all sorts of influences shape who we are. In addition, "normal" is a concept that eludes definition, even among psychiatrists and the most knowledgeable experts on human nature. It seems that human nature is still very much a work in progress.

A common objection to cognitive enhancement that Bailey doesn't address is the fear that human nature itself will change, that our individual and collective identity will be altered. But our sense of identity is already on shifting sand, and it's not at all clear what we might be losing when we choose to move forward with building an enhanced identity. If indeed we are somehow just what we are supposed to be, how do we assess the constant drive to become better? We would have to regard this drive as something unhealthy, perhaps even poisonous because it threatens to rob us of our true, and best, self.

Many bioconservatives hew to just such a view, and add that it is an insult to the Creator when we try to improve upon what He has created. Surely such an affront to the Creator's will should be accompanied by some spiritual or material disaster, a wage of sin, so to speak. I liken such views to the belief that humanity is the fully finished measure of all things. This is very much at odds with an equally long-cherished belief that striving for spiritual perfection is the chief concern of life, and it also conflicts with the concept of a fallen humanity guilty of original sin. Some creationists don't seem to be bothered by such inconsistencies. They regard man's participation in his own self-creation as hubris. They also foreclose any possibility that God may have intended all along that we are cocreators of ourselves. Humanity is not judged capable of making good choices that would uplift the individual and ultimately, the whole species. Nevertheless, a significant number of people already take drugs for neuroenhancement—reports estimate that anywhere from 4 to 25 percent of college students have used stimulant drugs to improve their academic performance, and people with intellectually demanding jobs are taking them in increasing numbers.[12]

As far as the distribution of neuroenhancing drugs, it's at the doctor's office where the rubber meets the road, since physicians are currently the gatekeepers of pharmaceuticals when healthy patients ask for stimulant drugs such as Ritalin, or memory-enhancing drugs such as Aricept. There are currently few clinical studies and few established guidelines to steer physicians who may be approached by patients wanting, not necessarily *needing,* neuroenhancement, whether through drugs or implants. Doctors at present are forced to rely upon FDA rules, which rest on a risk/benefit analysis. In 2009, the American Academy of Neurology (AAN) published recommendations for its members on how to respond to such requests.

The AAN guidelines, while affirming that the boundary between wellness and disease can be difficult to determine, even for medical experts, apply to those who have no diagnosed psychiatric or neurological disorder. The authors acknowledge that drugs such as Aricept can augment the memory of a person with normal, age-related memory loss and can improve executive function. The guidelines address whether a physician, just because he has been asked for a memory-enhancing drug, has an obligation to provide it. According to AAN, the asker becomes a patient—attendant with the usual rights and responsibilities—when the physician decides to treat him. The physician must navigate this critical juncture by either agreeing or refusing, and the decision of whether to enhance lies in his hands. Once the physician has agreed to treat, the asker becomes a patient, and then all the established rules of the physician-patient relationship apply.

The guidelines acknowledge that many medical procedures are currently available for purely cosmetic reasons, and it's not inherently unethical for doctors to provide such services. However, when seeking to guide the prescription of neuroenhancing medications, the most relevant guide is the FDA's policy toward off-label uses of drugs. Prescribing drugs off-label, or for purposes not strictly included in the FDA's recommendations, is widely accepted. Antidepressants are a class of drugs that are used for chronic pain, premenstrual syndrome, and several other conditions in addition to depression. FDA approval means that certain safety standards have been met, and the drug is within an acceptable risk/benefit range. AAN makes it clear that prescribing medications off-label is not contraindicated per se, but that the physician is not obligated to provide such prescriptions when asked. Beyond the need to educate patients as to the unknown effects of such drugs on a healthy person and the dictum to "do no harm," the doctor must rely on his best judgment when deciding whether to prescribe.

Making doctors the arbiters of neuroenhancement may make sense within the traditional framework of medicine, but because of the wider social effects of such drugs, the subject deserves a wider dialogue—one that includes representatives from the fields of science, medicine, ethics, sociology, politics, and even religion. Widespread use of neuroenhancement, either through drugs, implants, or even more advanced technologies, could have a dramatic impact on every facet of life, and it is urgent that we have a public dialogue about it. At the crux of this issue lie questions about freedom, personal choice, equality, responsibility, personhood, and citizenship. Abdicating such a wider dialogue, as has been the case so far, leaves the matter to be decided by the marketplace and private doctors, research-funding institutions like the National Institutes of Health and private foundations, and regulatory bodies like the FDA. While these institutions have arguably done a good job of directing traditional medical innovations, they are not equipped to grapple with the exponentially more powerful technologies brought about through converging technologies. Even the much-valued notion of patient autonomy has very little currency when patients have almost no understanding of the possibly powerful, unintended consequences of a radical new technology.

The burden of education to adequately inform patients is far beyond what the current system can provide. Human enhancement, being introduced as it is through the conventional pathway of medicine, should be an integral part of the public dialogue if concepts like personal choice are to have any meaning. Bioethics should be a mainstream concern, and should be taught as part of the main curriculum in all public high schools. Informing a patient about the ramifications of neuroenhancement can't be done satisfactorily by the doctor handing her a one-page, or even a multi-page, consent form during an office visit. The day is upon us when knowledge of cutting-edge medicine and its potential for human enhancement is a key ingredient if one is to be a truly informed citizen and an informed participant in a democracy.

One question that focuses the issue like no other is the question of whether to equip children with neuroenhancing treatments. While we tend to rely upon the rights of parents to make medical decisions for their children, the question of radical enhancement has wider social and ethical ramifications. The first question is to what extent can parents make intimate decisions for their children and at what age is a child allowed to make his own choices? Until neuroenhancement becomes uniform across society, to what extent can parents give their children competitive advantages over other children? What if a memory chip provided one child with all the answers to the questions on the SAT, while another child has to rely on her natural memory and many hours of study? Then we must take into account the dearth of research done on how various neuroenhancements might affect the developing brain over time. So far, stimulant drugs such as Ritalin have not engendered horror stories, but what if a fourteen-year-old chooses, over parental objections, to forego a brain implant that could influence her academic performance and thereby increase her chances of getting into a good college? If we are comfortable leaving the choice of whether a child receives psychotherapy or takes Ritalin in the hands of her parents, does that mean her parents can decide whether she receives a permanent brain implant that could change the course of her life?

British ethicist Hannah Maslen and a team of colleagues at the University of Oxford attempt to answer these questions in a 2014 article published in the journal *Frontiers in Human Neuroscience*.[13] They consider noninvasive brain stimulation techniques like TMS and tDCS as well as drug interventions in children. They point to a small body of research suggesting that there may be trade-offs in stimulating one region of the brain and not another. One study, published in 2013, found evidence that enhancing cognitive performance in one area led to poorer performance in another area.[14] If this is indeed the case, amplifying certain abilities at the expense of others in a growing child could abrogate her free choice to determine her own future later in life. While Maslen recognizes that parents can rightfully make *treatment* decisions for their children when a disease or deficit is present, we should hold decisions about enhancement to a different standard. The more a treatment falls into the enhancement category, the less prerogative is afforded to the parents and the longer a decision should be postponed until the child is able to decide for herself. Maslen leaves the threshold age for when a child is presumed mature enough to make decisions about her own enhancement somewhat open, suggesting that the threshold should not be before the child is sixteen. However, recent research has shown that the human brain may continue to develop well into the twenties and that judgment, in particular, is one of the last cognitive skills to develop. This seems to be an argument for postponing the right to make decisions to an even later age, but there is also the counterargument that beneficial interventions should be made at the earliest appropriate age, in order to have the greatest effect on the course of the child's life.

The lack of agreed-upon standards ensures that, at least in the near term, the risks and benefits of neuroenhancement will accrue to those who can afford it. For this reason, the most conservative thinkers suggest placing a moratorium on or even banning such technologies altogether. This is hardly realistic in a capitalist economy and a democratic society that thrives on competition. It's because of the short-term prospect that neuroenhancement will be unevenly distributed that such technologies should continue to be treated as medical treatments until such time as the larger society has reached some kind of consensus about their use. The regulation of brain enhancements under the current system, however inadequate, is better than no regulation whatsoever. Medicalizing things like a faulty memory, while it gives pause to some, allows the technology to be covered by health insurance, made accessible to a majority of people, and made amenable to longer-term research.

Many people will argue that making a radical new technology widely available will only invite unregulated experimentation by people who don't fully understand the risks. But people have been experimenting on their brains with mind-altering substances since the beginning of human history. There has never been a shortage of people willing to experiment through the use of recreational drugs and alcohol, and it's doubtful that everyone will have qualms about any accessible black-market treatment that offers them a better high or a competitive edge. Alcoholics and drug addicts who may be self-medicating for depression, anxiety, or some other psychiatric disorder may feel desperate enough to seek out a "back-alley" DBS treatment as well. The most dangerous alternative may be to not keep these treatments available and regulated through the medical system. If they remain so, all the checks and balances of the system, such as they are, can be brought to bear. Society can't prevent the truly determined from engaging in self-experimentation, but it can create a framework in which fewer people will feel compelled to seek enhancements outside of the care of a physician.

The brain is an amazing organ, but that is not to say that any brain is perfect. Diseases can affect any part of the body, including the brain. The fact that such phenomena as personality, free will, and spirituality are intimately affected by the brain does not call for a blackout on treating the brain when something goes awry, nor does it necessitate a ban on improving healthy brains to make them perform better. This discussion so far has been about the improvement of a neurological disease state and also the enhancement of general cognitive ability. But what about the affective, emotional brain and the psychiatric conditions that affect it? Mental illnesses such as depression, schizophrenia, and bipolar disorder are perhaps the most painful illnesses of all because they impact every aspect of life—self-esteem, relationships, sleep, energy, motivation, and the ability to be married or hold down a job, to name just a few. What good is a brain capable of vast computational ability if one is too depressed to endeavor to use it?

Mental illnesses such as obsessive-compulsive disorder (OCD) and treatment-resistant depression are now being treated experimentally with DBS. One researcher, Emory University neurologist Helen Mayberg, has identified a region in the cerebral cortex, the subgenal cingulate region, which she calls area 25, that she believes is overactive in severe depression that doesn't respond to drugs or therapy. She is now leading a clinical trial in which electrodes are implanted near the area that deliver mild pulses that actually down-regulate, or decrease, the activity in this region. Mayberg's study is the first hypothesis-driven study of depression and it seems to fly in the face of the way we have traditionally viewed depression. For centuries, depression was viewed as a condition in which several vital things—energy, attention, the ability to feel pleasure—are lacking. In Mayberg's hypothesis, area 25 is pathologically overactive.

I spoke with Mayberg about the latest clinical trial, and asked her about the "network" theory of mental illness that is prominent in her work. In fact, Mayberg and several colleagues published a paper on the use of DBS for depression way back in 2005 in the journal *Neuron*. In it, they write, "Converging clinical, biochemical, neuroimaging and postmortem evidence suggests that depression is unlikely to be a disease of a single brain region or neurotransmitter system. Rather, it is now generally viewed as a systems-level disorder affecting integrated pathways linking select cortical, subcortical and limbic sites and their related neurotransmitter and molecular mediations."[15] In our conversation, Mayberg stressed the fact that all the regions of the brain are interconnected, and that causing changes in one area is likely to initiate changes in the way all the networked areas function.

"All systems in the brain are networks," she said, "and advanced imaging is helping us see how different regions work together." She described area 25 as the intersection of a number of white-matter bundles and described the brain as a kind of power grid. "You have to understand the anatomy of the grid and find important intersections. DBS changes the whole choreography of a widespread system." Although the latest clinical trial is ongoing, so far results have been encouraging and suggest that quieting area 25 is working in most of the carefully selected patients, but so far the technology is still in its infancy. One thing Mayberg said suggests how dangerous DBS could be if not performed by a knowledgeable physician. A millimeter's difference in the placement of the electrodes "could make a difference between a partial response and a complete response." Imagine what could happen if one selected the wrong area or up-regulated a region that is already overactive. Mayberg said, "It wasn't random that we used area 25." She spent decades performing and analyzing PET scans of the brains of people with treatment-resistant depression, saying, "We studied this for twenty years before even going near a patient." She returned frequently to the theme that very little is actually known about the brain, that scientists have been trying drugs and other treatments with almost no up-front knowledge about how they might be working. They perform experiments that they hope will help, then develop theories based on reverse-engineering the results.

Mayberg's cautionary tone emphasized the potential dangers of brain-stimulating implants and even calls into question the use of milder techniques such as transcranial direct current stimulation. If tDCS does indeed affect brain circuits, then the correct placement of the headsets is critical. It's highly unlikely that the people using this technology know how to place them. The take-home message for any truly effective brain-changing technology should be "don't try this at home."

In the universe of future brain enhancements, DBS is one of the least radical prospects. At the far end of the spectrum, according to James Hughes, lies much more advanced brain-computer interfaces (BCIs) utilizing nanotechnology. It has been theorized that tiny, nano-sized robots (small enough to cross the blood-brain barrier) may soon be injected into the bloodstream. They would travel to the brain, where they would establish connections between individual neurons and external computers. This would allow a direct exchange of information in both directions, giving us unprecedented access to the Internet and to networking directly with other brains.[16] Some thinkers have even referred to a future "hive brain"—a true collective consciousness that we will all be able to plug into at will. To some, this represents an invaluable opportunity to expand one's consciousness to include vast amounts of information, including memories of the life experiences of countless other people. Others predict an unprecedented invasion of privacy when we give people and machines direct access to our memories, experiences, and thoughts, and perhaps receive irresistible commands to behave in certain ways. Fans of the television series *Star Trek: The Next Generation* will immediately think of the Borg, an alien race of biological beings who have melded so completely with their networked machines that no individual can think or act independently. The Borg "collective" dictates their every move and the collective has but one objective—to assimilate, through the forced placement of brain and body implants, every other race it encounters, or, if that's not possible, to destroy them. The Borgs' two mantras are "You will be assimilated" and "Resistance is futile." Once one's brain is invaded by Borg technology, the individual consciousness is destroyed, and resistance really is futile.

Although the story of the Borg is science fiction, the questions raised by this narrative are ones that we should be asking ourselves before we venture very far into BCIs beyond help for the paralyzed or the possible downloading of specialized knowledge. If humans can directly access a giant, two-way "Internet of human experience" in addition to the actual Internet, will people still be able to drive their own thoughts? Will the collective consciousness lead to such extreme conformity that individuality becomes nonexistent? What will happen to the concept of personal responsibility? Will the law preemptively punish people for thinking about committing a crime, and will our neighbors and coworkers be aware of our innermost thoughts? Bioconservatives are quick to paint doomsday scenarios, while optimistic futurists like Ray Kurzweil think we should embrace such developments because of their potential to broaden, enhance, and possibly empower the use of our brains.

The development of brain-computer interfaces will likely entail pressures, both social and economic, to use them, but democratic societies will most likely impose countervailing pressures to limit and control them, at least until much more is known about their impact on individuals and society. While we can easily celebrate the development of BCIs to assist the paralyzed, at this juncture we should be having a much wider debate about whether and how to place limits on BCIs. While augmenting human knowledge through direct downloading of computerized information may have some clear selling points, we also strongly value autonomy, privacy, and individuality. Will our commitment to these values prove strong enough to weather the onslaught of BCI technology? The concern is that we will rather easily accept, in the short run, interfaces that make life easier and more convenient, while being unaware of the long-term costs. If my bank provides an app through which I can move money between accounts using just my thoughts, such a convenience could be quite attractive. But what if the same technology gave my bank a window into my private thoughts? Where would data about my thoughts be stored, and who would have access to it? While most people would have objections to their bank, their employer, the government, and other people having access to their thoughts, the extreme complexity and constant time pressures of modern life make us highly vulnerable to any technology that offers us greater convenience.

I asked earlier what essential difference exists between having a technology like an app in a handheld device and having the same technology installed in our brains. BCIs illuminate this question by the prospect of directly plugging our brains into a universe of shared information, and plugging that universe, through a two-way street, directly into our brains. The difference between accessing the Internet on a handheld device is that our consciousness, privately enclosed within our brains, enables an active buffer of thoughts between us and the technology. While the technology might prompt, persuade, or even try to bully me into some behavior, my thoughts can intervene. I can still make decisions and judgment calls that may run counter to the technology. If these commands were actually received directly into my mind, would I recognize them as separate from my thoughts, and would I be able to resist them? Perhaps I would, but what level of effort would resistance entail? Isn't the very possibility that I wouldn't be able to resist if tired, unaware, or simply overwhelmed by too much information be enough to prevent me from accepting a BCI? When we lay the arguments out, it may seem obvious that most people would decline a technology that threatens to overwhelm their will, but the problem is that the arguments are not clearly laid out for the average person. These are the sorts of questions being considered mostly by science geeks, futurists, and philosophers, far from the mainstream of society, when they should be part of a mainstream debate.

Considering all the social and economic forces at work, it's not hard to see converging technologies moving seamlessly from the realm of medical treatments into the domain of enhancement.

The question is whether these technologies will then cross the line into a frightening level of social control and a life that none of us would freely choose if we were fully aware of the consequences.

Similar to the issue of deactivating implantable cardiac defibrillators when death is imminent, brain implants could have ghastly effects if not deactivated at the end of life. We must ask ourselves, given the dearth of information about what happens inside the dying brain, if we truly want to electrically stimulate it when a person is dying, and what sort of consciousness might be artificially prolonged—what gray area between life and death might we be inflicting on the patient? Yet deactivating brain implants is a prime issue that doctors will almost certainly shy away from, since deactivating a brain implant that is exerting some therapeutic effect could clearly be seen as harming the patient. How would we measure brain death if parts of a person's cognitive functions remain artificially stimulated while other parts of the brain die? It's only fairly recently that the medical community has been able to agree upon the criteria for brain death. Will we need to revisit the concept of brain death versus life support, and by what criteria will we define either?

In his 2005 book *The Singularity Is Near: When Humans Transcend Biology,* Ray Kurzweil writes of a time, the "Singularity," when computers will be powerful enough to exceed the computational power of the human brain. Predicted to occur in about 2050, humans are expected to have the ability, through BCIs, to completely download the memories, experiences, and thought patterns of their brains into a superpowerful computer at the time of death. Digitized information replicating the dying person's mind and personality could be housed in a robotic body or transferred to another person, thereby granting the deceased "immortality." But would a computerized copy of the brain at the time of death really be the actual person? Would it bestow continuity of consciousness, just as though the person had not died? What, if anything, would be lost in the translation?

These are at present unanswerable questions because the technology is so far merely theoretical. Futurists like Kurzweil and Kenneth Hayworth believe that mind uploading at death is a real possibility, and they see it as a great advance, solving the biggest problem of all—death. Of course, if such a possibility comes to fruition, not everyone will choose mind uploading. Those who believe in a nonmaterial component to the human person—a soul or spirit—and who also believe in a heavenly life after death would be less likely to settle for indefinite continuity of their life on earth. But for many people, the fear of death could make mind uploading irresistible. And there would be cases of parents of terminally ill children, for instance, who would be highly inclined to upload the minds of their children.

The question of mind uploading seems to rest on a resolution of the unsolved philosophical problem of mind versus brain. The materially inclined tend to believe that the mind is nothing more than the sum total of the electrochemical transactions occurring in the brain. In their view, once the brain stops working, the mind dissipates like an extinguished flame and no longer exists in any form. For others, the mind is much more than the sum of brain functions. It is something that is supported by the biological substrate of the brain but can exist outside of and separate from the brain. This debate has been going on for centuries, and in spite of advances in physics and neuroscience, it has yet to be resolved. Ironically, it may take experimentation in mind uploading to solve it.

The chief impediment to making mind uploading a reality is that very little is known about the mind to date. The second problem is one of converting biological information to digital information—a huge technological hurdle with many unknowns. The technique, as theorized so far, will be perceived by many as macabre, and as violating the integrity of the body to an unacceptable degree. So how, specifically, would mind uploading be achieved? Kenneth Hayworth, Ph.D., cofounder of the Brain Preservation Foundation (BPF) and a researcher working on advanced scanning techniques to create detailed "maps" of brain structures, described in a 2010 article how mind uploading might be done. Drawing from the article "Killed by Bad Philosophy: Why Brain Preservation Followed by Mind Uploading Is a Cure for Death," which is posted on the BPF website, the following is a summary of how the technique would occur. Fair warning: The description is not for the faint of heart.

A terminally ill woman checks into a hospital to receive a mind-uploading procedure. She is put to sleep with general anesthesia and her major blood vessels are hooked up to some external pumps. The blood is drained from her body as a poisonous chemical fixative, glutaraldehyde, is pumped through every artery, vein, and capillary. The chemical soon floods every cell in her body, including every neuron in her brain. The fixative bonds to the proteins in each cell, fixing them and preventing any decay. Then another poisonous chemical fixative is perfused throughout the circulatory system, this time preserving lipid, or fatty, molecules, which, along with proteins and carbohydrates, provide the chief structural components of the cells. Hayworth writes, "These two steps (fixation of proteins and fixation of lipids) are crucial in that they use chemical bonding to in effect 'glue' all of the molecular machinery within her cells together."[17] Needless to say, life will not return to this biological body. There's no turning back.

Next the vascular system is perfused with a heavy metal staining solution, such as uranyl acetate, to stain the brain and spinal cord for nanoresolution imaging. This is to allow the cell membranes to be visualized using an electron microscope. All the water must be removed from within and between cells and replaced with a plastic resin that will solidify tissues and prepare them to be sliced into tiny pieces and imaged. The tissues are made rigid by perfusing the body with ethanol, which draws out the water, to be followed by an organic solvent to leach out the ethanol. Increasing concentrations of plastic resin are circulated until "every region of intracellular [inside the cells] and extracellular [outside the cells] space is filled with pure plastic resin." Every cell, including every brain cell, is now inundated and fixed with plastic resin. The body has become a mixture of organic and inorganic, plasticized material, frozen down to the molecular level exactly as it existed at the moment of death.

The brain needs extra attention to assist with removing all the water and replacing it with plastic resin. Several holes are drilled into the skull and tubes are inserted to reach into the hollows, or ventricular spaces. These tubes push ethanol, solvent, and plastic resin into the dura mater sack (a leathery pouch enclosing the brain) while the same chemicals are being circulated throughout the circulatory system. Hayworth writes, "At the end of this process the patient's brain and spinal cord are floating in pure plastic resin and every nook and cranny of intracellular and extracellular space is also filled with this plastic resin."

The patient is then moved into a 60 degrees Celsius (140 degrees Fahrenheit) oven where the plastic resin in her brain and spinal cord are hardened into "a solid block." Hayworth continues, "Skin, muscle, vertebra, and skull bones are removed revealing the dura mater sack. This tough material is then peeled back to reveal a perfectly preserved brain and spinal cord (including initial segments of cranial and spinal nerves) encased in an amber-colored transparent plastic sheath. Every neuron, every synapse, every delicate neuronal process is now perfectly preserved down to the nanometer level—the most perfectly preserved fossil imaginable."

The solid, plasticized brain and spinal cord are next sliced into long thin strips, just 100 microns, or .01 centimeters thick, which are wound onto spools like tape. The result will be thousands of spools of brain strips, and these will be loaded into thousands of electron microscope scanning machines. The machines will scan the surfaces of the strips with almost unimaginably tiny electron beams, producing a high-resolution, two-dimensional picture. A focused ion beam is then used to burn away the top five nanometers of the brain strip, revealing a new surface to be scanned by electron beams. This process is repeated twenty thousand times, until the entire 100-micron-thick strip has been imaged in five-nanometer increments. Burning down in five-nanometer increments adds a third dimension to the image. Needless to say, the strips are completely destroyed, and in their place we have a 5×5×5-nanometer-resolution image of them in a digitized format.

Next the three-dimensional, ultra-high-resolution image of the brain and spinal cord are fed to a computer that maps out every neuron and synapse, estimating the type and strength of every connection. Since Hayworth estimates this procedure is taking place one hundred years from now, there is still room for interpretation of the data based on the previous one hundred years of advances in neuroscience. Also preserved were the cranial and spinal nerves that connected the biological brain with the body's nerves and muscles. The computer simulation of the patient's brain and nervous system are interfaced with a robot body—one that the patient can manipulate with her thoughts as her own—and voila, immortality has occurred. Or has it?

There is more than one school of thought that might be

applied to the question of whether a digitized simulation of our brain is actually us, but until an attempt is made to actually do it, all we can do is speculate. Those who believe that there is some vital essence to a human being that is more than material will regard the attempt to duplicate a person in such a way as not only futile but grotesque.

We must recognize that we know far too little about the mind or the brain to ascertain all that it would take to replicate a person. If mind uploading is ever attempted, a facsimile of the mind will be created, but a digitized copy is not an original, no matter how detailed the copy. The mind is characterized not only by its physical structures and electrochemical reactions, but by the movement of mental processes which, it can be argued, are driven by, among other things, our will. How would the phenomenon of volition be approximated in a computer simulation? Some manifestation of will, or some effective substitute for it, would be needed for us to direct our attention and to make decisions and act on them. In addition, living in our biological bodies, we are not only aware, but self-aware. How would a computer simulation create awareness of a self that performs the executive functions of the brain? Through computational programming, we might be able to get our robot facsimiles to do all sorts of things, but none of those things would be done with free will as we understand it.

There are many facets of the mind about which we currently have no way of knowing if we were reproducing with our facsimile. Some of these could be integral to making us who we are. They include a subconscious mind, emotions (which are interconnected with a sentient body), the ability to make value judgments, and a potential soul or spirit in addition to self-awareness. People can and do differ in their opinions about these phenomena, but even if we assume that only one of them is valid and all the rest are just invalid theories, we have no idea how the absence of that one element would affect a would-be human being. The lack of only one or more of these phenomena could mean that our robot is but a macabre reminder of the person we have lost.

But let's assume for the sake of discussion that mind uploading turns out to be a smashing success. The procedure, once perfected, truly allows for continuation of our subjective consciousness and as far as we or anyone around us is concerned, we have never died. We and our loved ones have had to make adjustments to our personality being domiciled in a robot body, but over time everyone has gotten comfortable with the new us. If this happens, not everyone will choose to escape death, but it's likely that large numbers of people will choose mind uploading. This means that society will have a number of serious problems to solve.

Overpopulation, which is almost certain to be a big problem in the coming years, would go into warp drive if a large percentage of the population decided to live forever. People would be obliged to have fewer and fewer children to accommodate the diminishing resources of an already overcrowded planet. The age composition of the population would change, with more and more extremely old people (though presumably with a succession of high-functioning, up-to-date robotic bodies) and there would be less room and fewer resources for young people and their children. Assuming that radically extended biological life spans comes to fruition, society will already have a high percentage of older people, leaving even less room for children and young people. Governments would have to mandate fewer children or even ban reproduction, violating what most people regard as their right to have and raise children.

Today we value the energy, fresh perspectives, and new

ideas that young people contribute to society, and mass mind uploading would severely limit the number of young people able to make such contributions. It may be possible for those who live forever to bring continuous innovation to culture and society, but we can only speculate about what type of society extremely old people would create. It may be that young people and children come to be cherished much more than they are today because of their rarity, but there might also be a backlash against them because planetary resources will be so scarce. One thing we can be sure of is that the imperative to control the size of the population will lead to a far different society than the one we live in today.

Another issue that will need resolving is that by the time many people become terminally ill, a significant number of them will already have extensive brain changes like the plaques and tangles seen in Alzheimer's disease. It would be convenient to assume that by the time mind uploading becomes technologically possible, there will be a cure or a prevention for Alzheimer's, Parkinson's, epilepsy, stroke, and other neurological conditions, but we can't make such an assumption. Consider for a moment that there is a period when we have good treatments, but no cure for any of these conditions. Copying the brain of an Alzheimer's patient would likely reproduce the disease. Does this mean that we would deny mind uploading to those whose brains don't meet a certain threshold for health? It may be that once we have the brain in a digital format, we can tweak it to correct for diseases, but then we would have a less faithful copy of the person at the time of death. Since changing one area of the brain can result in many more changes to the neurological network, making the necessary corrections to cure a disease might have unintended consequences that may or may not be reversible.

A major ethical stumbling block will be the decision of when to perform mind uploading, even in a terminally ill person. As described above, someone—preferably the patient himself—must make a decision about when the procedure can take place, given that it kills the biological body. One might say that it should be done when a person is actively dying, but this argument inevitably meets with problems. First of all, the level of technology needed means that the procedure could only be done in the right type of facility, and optimally it would be scheduled so that the necessary personnel and equipment would be available. This would be no help to a person who dies instantly in a car crash, for example, far from a major medical facility. It could be many hours before a body reaches a facility that can perform the procedure, and by that time, significant changes could have occurred in the brain. The brain may already be too changed to make a functioning copy. Due to the uncertain nature of when and how death occurs, some people won't be able to take advantage of mind uploading even if they wish to.

But let's assume that optimal conditions exist. The patient is either in the hospital or in a hospice facility, and his caregivers agree that his life is coming to a close. Say the patient has indicated that he's ready to go forward with mind uploading, regardless of the knowledge that the procedure will irreversibly destroy his biological body. How will doctors, nurses, and technicians know that they are not cutting his life too short? Even the most experienced doctors currently find it difficult if not impossible to say for sure when a person is actively dying, and the need to schedule a mind uploading means that the procedure must take place at some time before natural, biological death occurs. Doctors may estimate that the patient has only hours or days to live, when the patient in fact could survive for several weeks or months. The situation is not like one in which a patient reaches brain death while on life support and a decision must be made to stop artificial respiration and the heartbeat. In this case, the patient is not dead, but the mind uploading will certainly kill him. Should something go wrong with the procedure, he is not coming back, nor will he have a few more months with his loved ones.

Legally, if a person's brain is alive, he is alive. To upload the brain, someone must decide to end biological life—in other words, to kill the patient. How would this square with the traditional healing role of the physician and the dictum to do no harm? If doctors are reluctant to deactivate cardiac defibrillators even when they cause suffering at the end of life, how will they be persuaded to poison someone to death with chemicals and perform the radical assault on the body that mind uploading entails? We'd like to assume that the philosophical question of whether a digital facsimile of a person's brain constitutes immortality will be resolved by this time, but that is unlikely to be the case. Technology moves much faster than the advancement of social and cultural values. It's likely that a controversy will exist about mind uploading that will be hard to untangle, and there may never be a complete social consensus about it.

Then there's the question of a patient's fitness to decide if and when she wants mind uploading at all. Given the current reluctance of people to create advance directives (a 2013 survey by FindLaw found that only about a third of Americans have a living will)[18] and the spotty record of physicians in complying with them, most end-of-life decisions are made during crisis situations. They are rarely made in advance by younger people. Americans don't like to deal with issues related to death, and this is an impediment to having clearheaded decisions in place when a crisis strikes. In addition, even if medical power of attorney has been given to a loved one, grieving family members can make irrational decisions.

Imagine for a moment that a fifty-year-old woman with no advance directive suddenly has a severe stroke that damages her brain to the point that she is in a semivegetative state and is paralyzed for life. She will have missed the boat to having her normal mind uploaded even if that had been her stated wish. Being confined to a bed, she becomes vulnerable to pneumonia and the other infections that often take the lives of people who are paralyzed. A few years later, she develops an antibiotic-resistant infection and soon she is actively dying. Her husband believes that her wish under such circumstances would have been to pass away peacefully, but her parents insist on uploading her mind, notwithstanding her severe cognitive and physical disabilities. They feel that not uploading her mind would be tantamount to murder and make the case that even in her disabled state, her life has value.

Both the husband and the parents approach the woman's doctor, asking that their wishes be honored. Now the doctor and the hospital are in a legal predicament. The husband, who wants natural events to run their course, threatens to sue the doctor if he hastens his wife's biological death through mind uploading, which would subject her to the prospect of living forever in a highly disabled state. The parents, meanwhile, threaten to press charges of murder against the doctor and to sue the hospital for discrimination against a disabled person if it refuses to upload the mind of their daughter. The case is taken to a judge. The judge hears the arguments on both sides and decides that since the woman didn't have an advance directive, the decision should lean in the direction of "saving" her life and giving her every chance to avail herself of possible future treatments that might correct her disabilities. He orders the hospital to upload the woman's mind.

The woman's mind is uploaded, and the husband is left with a robot with a mind that in no way resembles his wife before the stroke. Although various technological adjustments are made to help the robot move its body, the mind is still severely disabled. The husband is now expected to care for this robot for the rest of his life, and if he chooses mind uploading, that could quite possibly be forever. He takes his wife's parents, doctor, and hospital to court, suing for a lifetime of damages, including emotional pain and suffering. What he really wants is to shut down the robot that is nothing more than a painful reminder of what happened to his wife and move on with his life. However, the law recognizes the robot as a full-fledged person and deems that turning it off would be murder.

Those who remember it will see a strong resemblance to the high-profile case of Terri Schiavo in the mid-2000s. While I have simply extrapolated possible developments based on our handling of such issues today, it's likely that there will be other bioethical quandaries in the future that we have yet to even imagine. Other scenarios come easily to mind. What if mind uploading did not create continuity of consciousness but did create a highly complex and high-functioning robot with some, but not all, of the attributes of humanity? What rights would these robots have, and what role would they play in society? Would everyone have a right to mind uploading? What about convicted murderers and other antisocial personalities? What happens if we want to shut dysfunctional robots down? What if the robots themselves want to be shut down? One could conceivably ask a great many more questions, but given that we have yet to solve so many of our current pressing problems, we have a long way to go before we can even begin to answer them.

# 7

## The Ageless Society

The fight against aging is big business in today's youth-obsessed society. Americans spend about $80 billion a year on products and treatments that promise them smoother skin, thicker hair, slender, muscular bodies, and agile minds. These products include dietary supplements, hormone replacement therapies, expensive skin care products, and cosmetic surgery, to name just a few. The baby boomers, who immortalized the slogan "Never trust anyone over thirty," are now entering their sixties, and by all indications, they don't intend to go down the road of old age without a fight. We're already seeing one effect of a graying population. In the past decade, there has been a 77 percent increase in the number of cosmetic surgeries performed. In 2014, there were 15.6 million of them, and the figure continues to climb.[1]

Americans would do well to be skeptical of over-the-counter therapies that claim to reverse aging. Most antiaging therapies, such as special diets, nutritional supplements, and cosmeceuticals (special creams and serums that claim to slow or reverse skin aging), are completely unregulated and don't have to be proven effective in order to go on the market. Few have been tested with scientific rigor, and it's anyone's guess whether they make any difference. That, however, hasn't stopped millions of people from pursuing them avidly.

In a field traditionally rife with exaggerated claims and even outright quackery, the administering of human growth hormone shots is one of the most expensive and controversial antiaging treatments. The cost can exceed fifteen thousand dollars per year, and among the possible side effects is a higher risk of cancer. The dietary supplement DHEA, claimed to be a precursor to the hormones estrogen and testosterone, is sold online for about thirteen dollars per capsule. People are willing to pay high prices and take all kinds of risks with their health with hormones, even though the National Institute on Aging recently told the Associated Press, "The truth is that, to date, no research has shown that hormone replacement drugs add years to life or prevent age-related frailty."[2]

Despite the dubiousness of antiaging products throughout history, we are about to enter a new era, one based on decades of research. There is a legitimate, and growing, field in science that is testing a number of potential interventions in the aging process itself that have already shown significant success in animals. A new culture is taking hold in the scientific community that clearly approaches aging as a disease and, rather than just chipping away at age-related diseases, some scientists are searching for a "cure" for aging itself.

Many gerontologists still maintain, as does S. Jay Olshansky, that "aging is not a disease any more than puberty or menopause are."[3] But legitimate scientists have started to attack the aging process based on the premise that aging *is* a disease that is amenable to treatment, if not to an outright cure. How we define aging will probably continue to be debated for some time, especially among bioethicists, but real antiaging science is already illuminating the debate with an infusion of carefully studied scientific evidence. The latest science through stem cell therapies, genetic interventions, nanotechnology, and antiaging drugs could be game changers when it comes to how we view aging and mortality. Fifty years from now conventional wisdom regarding aging and the human life span is likely to be dramatically different and exponentially more informed than it is today. The mists surrounding an ancient problem are starting to clear.

In chapter six, I raised a few negative "what ifs" about indefinite life extension based on mind uploading. In this chapter, I'll discuss the brightening prospect of greatly extending our life spans and, more importantly, our health spans, in the bodies we were born with. There's no question that all industrialized nations and quite a few developing ones are facing an aging revolution unlike any seen in history. In 1900, the average American life expectancy was 47, and today, according to the U.S. Centers for Disease Control and Prevention, it's 78.8 years. Given the scientific tipping point we're approaching, it's probable that we are in for even more dramatic increases in the twenty-first century.

Not everyone is on board with welcoming radical life extension, and arguments about why we shouldn't tamper with the life span abound. But most of the arguments against life extension rest on current stereotypical images of old age as a period of disability and decrepitude. Our reference points seem to be confined to past-generation realities of a less medically sophisticated era when the average person was far less informed about exercise, diet, and other lifestyle factors that extend not only life, but health. The stereotypes of aging that we hold in our minds today have almost nothing to do with what aging will look like twenty, thirty, or fifty years from now.

Antiaging research so far has been primarily focused on extending the life spans of the humble fruit fly, worm, mouse, and yeast, organisms that share some of our genes and have naturally short life spans. The short life spans allow researchers to easily study gains in longevity, and significant gains have been made through various techniques. Although transferring accomplishments in other species to humans is always a tricky proposition, nature holds a treasure trove of information about the mechanisms of aging that could benefit mankind. Through the study of other species, scientists are learning about the molecular and metabolic processes of aging and finding numerous ways to manipulate them.

So far the best known strategy to extend life span in mammals by as much as 45 percent is extreme calorie restriction, which seems to put cells into a kind of hibernation state that postpones a stage called senescence, when cells stop dividing, degenerate, and eventually die. But for humans to achieve significant life extension this way, they would have to limit their diets to the point of constant hunger—not an option for most people in our "super-size-me" culture. There have recently been other gains in antiaging research that could translate into more realistic interventions in the aging process. While no one has yet found the mythical Fountain of Youth, there are many promising avenues and tantalizing accomplishments in antiaging science.

Researchers at the Buck Institute, an antiaging research organization in the San Francisco area, have found ways to manipulate the genes in worms in a way that quintuples their life span, raising the question of whether humans have similar genes that might be used to radically extend life.[4] Another intriguing development occurred in 2014, when several separate research teams—at Harvard, Stanford, and the University of California at San Francisco—discovered that when the blood of adolescent mice was infused into old mice, the old mice were rejuvenated. Now their task is to uncover which properties in the blood of young organisms might be isolated and introduced into the blood of older ones. In other work reported in 2010, researchers were able to dramatically reverse aging in mice by reactivating an enzyme called telomerase. The work confirmed what many antiaging researchers have hypothesized, that by activating the enzyme that keeps telomeres—the "caps" on the ends of chromosomes that get shorter each time a cell divides—intact, cells could keep behaving like youthful cells indefinitely. Telomerase is turned off in adult humans, and the risks of turning it back on are not known, but the enzyme seems to do exactly what scientists have long theorized it would do.[5]

Another target of research is the above-mentioned senescent cells. When cells become senescent, they send out distress signals that induce inflammation, a process widely associated with arthritis, heart disease, Alzheimer's disease, and a host of age-related conditions. Researchers at the Buck Institute are currently investigating a drug that it is hoped will shut down the distress signals coming from senescent cells and thereby deactivate their toxicity to other cells.[6]

Stem cell research continues to be one of the most promising areas in the science of reversing aging, whether it's by selectively regenerating and transplanting new cells, tissues, and organs, or by rejuvenating cells inside the body itself. Through recent breakthroughs in reprogramming adult cells, it has been proven that cells from even very old people can be turned back into young, vital, energetic cells. Several scientific teams throughout the world have succeeded in taking adult cells and, through the introduction of various genetic factors in culture, induced those cells to return to undifferentiated, "pluripotent" stem cells that can become a brand-new, youthful cell of any type. In 2011, a French team even cultured pluripotent cells from a one-hundred-year-old donor.[7] The ability to turn adult cells, even very old ones, into pluripotent cells is paving the way for the ability to grow new tissues and organs that are perfectly genetically matched to the cell donor. These organs should show no signs of aging and last a lifetime without being rejected by the body. They could literally add decades to a person's life. But could scientists someday know how to coax our bodies into producing more adult stem cells that would heal organs from within? A large body of research suggests that this is indeed possible within our lifetime.

Another very promising area for research in life extension is what some call synthetic biology, or manipulating the human genome through various tactics gleaned from studying aging in other species. A team led by NIH researcher Toren Finkel reported in 2013 that it had increased the life span of mice by 20 percent (the equivalent of about sixteen years in a human) by inhibiting the expression of a single gene.[8] There are probably multiple genes involved in human aging that could be subject to manipulation, and many clues in nature about how to extend youth and life. By studying nature, scientists are amassing many clues to how aging takes place (or doesn't) in other species, and the processes uncovered could potentially translate into human antiaging therapies.

Some species, such as the cactus-like aquatic creatures called hydras, don't age at all, and a species of jellyfish can actually revert from a mature state to an infantile state, escaping natural death altogether until killed by a predator or an accident.[9] Scientists are studying these creatures to unlock the biological mechanisms associated with youthfulness and longevity, hoping to modify the human genome to mimic the strategies used by these organisms, a process that Ed Regis and George Church refer to as "genetic data mining." When life- and youth-extending genes are isolated in other organisms, such as hydras, the next step will be to take snippets of their DNA and insert it into human cells, where it becomes part of an individual's genome. This might sound bizarre, but scientists have already succeeded in altering the genomes of certain animals by inserting the genes of other species. Brightly glowing dogs, monkeys, cats, rabbits, and many other species have been genetically engineered to express jellyfish genes that make the animals fluorescent in darkness.

The splicing of genes from one species into another so far appears to be safe, but it comes with a "yuck factor" that will turn many people off. However, we should keep in mind that this science is in its infancy. The goal is to isolate very specific genetically driven processes and find ways to introduce those processes into the human body. If scientists can tease out what prevents hydras from aging, there may be a safe way to initiate that process in people, by inserting specific genes into humans, or through some other method. Humans that come to harbor the genes of other species will become what scientists term "chimeras," and we can expect bioconservatives to have major reservations about the practice. It's possible that future humanity will possess not only numerous artificial body parts, but genetic material taken from other species, making it even more difficult to define "human," since humanity will no longer be characterized by a purely human genome. On the other hand, once scientists have identified desirable molecular processes, they may be able to re-create them in humans through drugs or other interventions that don't require changing the genome. The staggering variety of vegetable and animal life on our planet may harbor untold genetic richness that could end up benefitting humanity—just one more pressing reason to protect the natural environment and to prevent species extinctions.

Genetic manipulations, blood transfusions, and stem cell treatments are all rather elaborate solutions to aging. One might ask, "Why can't we just find a drug that retards aging?" It so happens that in the next few years, we may all have access to a bona fide antiaging drug. Several antiaging drugs are now being tested. The best known one, being developed by the Swiss pharmaceutical giant Novartis, is a derivative of a bacterium isolated from a soil sample taken from (of all places) beneath one of the statues on Easter Island. The derivative, discovered by the Pakistani-American researcher Suren Sehgal, is called rapamycin.

In the 1990s, Sehgal found that rapamycin not only had antifungal properties but also suppressed the immune system, and in 1999 the U.S. Food and Drug Administration approved its use in transplant patients. Since then, more and more uses of rapamycin have come to light. It has now been approved for the treatment of certain kidney, lung, and breast cancers, and it's routinely used as a coating on cardiac stents to prevent scarring and further atherosclerotic buildup. In the last ten years, it has been discovered that the drug delays the onset of many age-related diseases such as heart disease, cancer, and Alzheimer's, and it also seems to prevent the signs of normal aging. Although the work demonstrating that rapamycin prevents aging was done in mice, the way it works is almost too good to be true. The drug was shown to prevent age-related changes in "multiple different organ systems," suggesting that it affects the global process of aging in a fundamental way.[10] Male mice that received rapamycin lived 9 percent longer than mice that didn't, and female mice lived 14 percent longer. While that is by no means immortality, to translate it into human years, a sixty-year-old woman taking the drug could expect to live to age ninety-five.

Rapamycin creates molecular conditions in the body that resemble what happens during calorie restriction—the best known, surefire way to postpone aging in mammals. The rapamycin molecule down-regulates a cellular pathway that controls metabolism and growth. The drug tricks this pathway into a partial hibernation state so that the cells, rather than dividing and gobbling up energy and nutrients, start to clean up debris from past divisions, recycle old proteins, and conserve energy, just as they would if they were in a state of calorie restriction. The result is retarded aging without the hunger pangs. In large doses, there is a downside, though: rapamycin suppresses the immune system, something you don't want to do in older adults, whose immune systems are already compromised. This could have been a deal breaker for Novartis to continue to develop the drug as an antiaging medication, but on Christmas Eve in 2014, a watershed paper was published in the journal *Science Translational Medicine*. The paper presented findings from human trials conducted in Australia and New Zealand and found that everolimus, a derivative of rapamycin, actually *improved* the immune response of elderly volunteers when given in small doses.[11] Rapamycin's virtues continue to emerge in mouse studies that scientists hope will translate to human therapies. In mice, the drug has been found to reverse cardiac aging, reduce age-related bone loss, shut down the chronic inflammation associated with aging, and even reverse Alzheimer's disease, something no other drug has been able to do.

As of this writing, hopes are riding high on rapamycin. Novartis is pushing ahead with further testing, and a new study is examining the use of rapamycin in pet dogs. It's not clear whether the FDA will ever approve the drug as strictly an antiaging medication. The FDA doesn't classify aging as a disease. But it's hard to imagine a drug that effectively postpones aging not being made widely available. It's possible that anyone with early signs of an age-related condition will be able to obtain it with insurance coverage. And the wealthy, of course, will be able to buy it without insurance coverage.

There are other antiaging drugs in the pipeline, including

the diabetes drug metformin, which has also extended the lives of mice. Other drugs are targeted at specific age-related changes. The drug bimagrumab targets the muscle loss and general frailty that comes with old age, and it is now undergoing clinical trials. Other drugs are aimed at processes like the loss of cartilage in knees and other joints. Then there is a new class of drugs called senolytics, which destroy senescent cells and rid the body of the distress signals that accelerate aging. One, already on the market, is the anticancer drug dasatinib, which is sold under the brand name Sprycel.[12]

It's possible that many of us will soon be on a cocktail of drugs that can extend our healthy life span, or health span, by 10 to 20 percent. That may not seem too impressive, until you consider the yet-to-be-known benefits of an exponential rise in health consciousness among current generations. We are just beginning to see what additional declines in chronic, degenerative diseases of aging will be brought about by the adoption of today's young and middle-aged people of not smoking, eating healthy diets, and exercising regularly over many years. These factors could precipitate a notable jump in the average age of death that we can't accurately predict until these generations reach old age. In addition, no one knows what the future longevity dividend may be of huge numbers of people doing things like starting cholesterol-lowering medications in middle age. The grandparents of today's baby boomers lived longer than their parents, even though a great many of them smoked, ate high-fat, sugary diets, and got virtually no exercise. It seems that today we are in the midst of a "perfect storm" of life and health extension, and today's health-conscious generations may even reach an age that surpasses current predictions.

There is one dark cloud on the horizon. The main impediment to long life today is obesity, and it must be recognized that the obesity epidemic will shorten the lives of millions of people. We know now that the lifestyle choices we make throughout life have a tangible effect on how well we age and how soon we die, and being overweight accelerates age-related changes in many ways. Fortunately, there are potentially effective treatments for obesity under development.

As significant as the gene therapies are to eliminate obesity, as discussed in chapter one, such interventions are only the tip of the iceberg in antiaging science. In terms of the long-term potential to radically extend the human health span, we have barely begun to take baby steps. Most experts would say that the biggest advantage to embracing antiaging drugs and other treatments today will be to live long enough to see the next generation of technologies that will possibly extend our lives not by years but by centuries. Ray Kurzweil cowrote a book (published in 2005) titled *Fantastic Voyage: Live Long Enough to Live Forever,* setting a goal that he and others believe is within the grasp of many people alive today. According to Kurzweil, once the biotechnology and nanotechnology revolutions really take hold, humans could live for many centuries. Not everyone believes this is achievable, and not everyone who believes it's achievable thinks it's a good idea, but the idea of very long health spans is quickly gaining ground.

Aubrey de Grey is perhaps the world's best-known advocate for antiaging medicine. The outspoken British gerontological researcher with the ponytail and the long beard is completely unabashed in his view of aging as a disease. De Grey thinks it's unproductive to focus on the diseases of aging rather than the aging process itself, which causes them. In a 2014 interview, he told Gian Volpicelli for the online magazine *Motherboard,* "diseases of old age are not really diseases. They are aspects of aging, side effects of being alive. If you want to cure them, what you'll have to cure is to be alive in the first place."[13] What de Grey means by this is that much, if not all, of the effects of normal aging are brought about as byproducts of normal human metabolism. It's well known that cellular metabolism inevitably creates toxic molecules and misfolded protein fragments that gunk up the body's cells and tissues, and that over time the body gets less and less efficient at clearing them away. Oxygen free radicals are the best known culprits, but there are many others. However, there's more to aging than just toxic cellular debris. De Grey founded the SENS Research Foundation to promote research into the mechanisms of aging and outlined his theory of aging in the 2007 book *Ending Aging*. SENS stands for Strategies for Engineered Negligible Senescence, and de Grey has identified what he believes are the seven causes of aging, all of which are potentially amenable to intervention. His theory of aging is not uncontroversial, and he hasn't explicitly proven that it's accurate, but it outlines what he and some other scientists recognize as important research targets. De Grey's seven causes of aging include: extracellular junk (protein fragments and other debris between cells); cell senescence (the wearing out and death of cells along with the inflammation caused by "sick" senescent cells); extracellular crosslinking (too many and inappropriate bonding between special proteins that link cells together, making tissues stiff and brittle); intracellular junk (oxygen free radicals and other debris inside cells); mitochondrial mutations (mutations or mistakes in the genetic material outside a cell's nucleus); cancer-causing nuclear mutations (mutations or mistakes in a cell's nuclear DNA); and cell loss, which leads to tissue atrophy.

De Grey offers theoretical interventions for each of his causes of aging, and is emphatic in his claim that aging itself is not simply correlated with age-related diseases; it's the actual cause of them. He believes that in order to arrest aging, one would have to address most if not all of the causes. Intervening in one or two of them might have a limited effect, but if interventions were found for all seven of them, people could remain biologically twenty-five for a thousand years or more.[14] In 2005, de Grey predicted that we were about twenty-five years away from radical life extension. Only time will tell if this prediction comes true, but there are good reasons to be optimistic about antiaging science in general.

The science of antiaging, long considered a trail of broken dreams, is going mainstream, with the baby boomer market of seventy-seven million consumers almost guaranteed to make such efforts lucrative.[15] In the years since de Grey first made his predictions, major pharmaceutical giants like Novartis and Glaxo-SmithKline have started investing real money into antiaging medicine. Now the behemoth technology company Google has gotten into the game. In 2014, it launched Calico Labs and partnered with the pharmaceutical company AbbVie on a $500 million venture focused on human longevity and antiaging research. In fact, population pressures such as the worldwide aging of the population promise to send antiaging research into warp drive. It's not only private industry that has an interest in finding therapies that fight aging; governments have a vested interest in keeping older people healthy and working as long as possible to avoid both skyrocketing health care costs and the cost of funding millions of pensions soon to become due.

To truly understand the longevity revolution that has been set into motion and what it means to society, we need to forget almost everything we know about aging. Visions of frail, chronically ill elders sitting in rocking chairs, lost in memories of bygone days, is a stereotype that bears little to no resemblance to the experience of aging that most of us will know. This would already be true even if no additional medical innovations occurred in our lifetime. But the past few decades of research have created a momentum, especially now that technologies have started to converge, that will culminate in a new paradigm: that of successfully maintaining health and youth rather than battling diseases one by one.

The most dramatic advances are likely to come from the field of nanomedicine, or the medical application of nanotechnology. Nanomedicine is a quantum leap more advanced than conventional medicine because it works at the molecular level, where atoms interact with molecules like protein, DNA, enzymes, amino acids—the very machinery of our cells. We are talking about controlling and manipulating molecules that are one one-thousandth to one ten-thousandth the width of a human hair, in a field that is part chemistry, part biology, part physics, part computing, and part robotics. Nano-sized drugs and robots can easily penetrate cells and tissues, where they can home in on the tiniest targets. K. Eric Drexler, who wrote the groundbreaking 1986 book *Engines of Creation: The Coming Era of Nanotechnology*, observed that "physicians using scalpels and drugs can no more repair cells than someone using only a pickax and a can of oil can repair a fine watch."[16] This is a fundamental limit to conventional medicine given that the parts of us that go haywire in disease and aging are the cells themselves. Even most drugs, when introduced to the body, simply tumble aimlessly through our bodies until they bump into a certain molecular receptor and click into place. Only through this haphazard process can they have any effect. Add to that the reality that, for most of our drugs, doctors don't know how they work—they just know that sometimes they do.

Through nanotechnology, scientists are now designing special molecules that are precisely targeted at carefully identified molecules in the body. We're emerging from an era of medical hit or miss to an era of consciously designed therapies, one that promises to be much more effective than the clumsy and only partially successful era of conventional modern medicine.

A frequent criticism of biomedical researchers doing antiaging research is that some of them tend to reduce the search for medical cures to engineering problems. Basic medical research concerns itself with identifying the causes of disease, whereas the engineering approach would concern itself with only achieving a desired end, such as the correction of some abnormal process. But this approach, promoted by chemical and bioengineers, computer programmers, mathematicians, and experts in artificial intelligence, may not be so far off base. The most radical theories in antiaging might well benefit from an engineering approach. I'm referring to the possibility, proposed by Drexler and others, that in the foreseeable future we'll have injections of miniscule nanorobots that will patrol our bodies, seeking out cancers, DNA mutations, excessive crosslinking of proteins, and a host of other malfunctions. These bots will be programmed with a blueprint of what a healthy cell looks like structurally, and they will destroy "unfixable" cells like cancer cells and fix cells that can be saved, returning them to a healthy, youthful condition. They might very well be able to address most if not all of Aubrey de Grey's theoretical causes of aging.

By returning our cells to a youthful condition, we will be repairing tissues, organs, and our entire bodies to a youthful, healthy state. Rather than waiting until the body suffers disease and aging and then trying to address the problem after the fact, we will periodically receive infusions of the nanobots, which will repair any age-related damage, keeping us biologically young indefinitely. The bots themselves, after their work is done, will simply dissolve, be broken down, and have their byproducts excreted in the urine. Diseases can be the result of a massive cascade of malfunctions. It seems far simpler and more achievable to maintain the body in a "happy" state of molecular and cellular health than to try to reverse damage that might consist of a thousand chain reactions. And while patients are by no means alike, the healthy cells of our body all contain the same DNA. Nanobots, because of their infinitesimal size, will be able to snip out damaged sequences of DNA, for example, and rebuild the sequence based on a healthy genetic blueprint of what our cells are supposed to look like.

This technique is startling, but it is by no means magic. The principle by which nanomedicine works is actually pretty simple. It involves breaking the task of cellular repair down to its tiniest parts and methodically addressing them one by one.

Drexler goes on to name the known biological processes that nature has already proved possible that nanobots can mimic. The first process is access to cells and tissues, and this is proven possible by the process of white blood cells leaving the bloodstream and entering tissues. Viruses further prove this process because of their ability to enter cells and do damage; nanobots will do just the opposite. Second, nanobots will recognize molecular targets by touch, something that antibodies in the immune system already do. Nanobots will disassemble faulty genes, amino acid chains, and other "broken" machinery, just as digestive enzymes break down the food we eat. Then nanobots will rebuild molecules just as cells do when they replicate, and they will reassemble cells just as nature does when the cells divide. In addition,

> By working along molecule by molecule and structure by structure, repair machines will be able to repair whole cells. By working along cell by cell and tissue by tissue, they (aided by larger devices if need be) will be

able to repair whole organs. By working through a person organ by organ, they will restore health.[17]

The larger devices Drexler refers to are external computers. The nanobots themselves will include tiny, programmable computers that will be powered by naturally occurring chemicals already in the body's cells. Occasionally, the bots will need to access more information than can be stored in their onboard computers, so they'll communicate wirelessly with larger computers.

The homing process of nanobots is also rather deceptively simple. The bots will read the DNA blueprint from our healthy cells, then compare groups of cells to find aberrations. They will home in on the aberrations, break them down, then rebuild the molecular machinery in keeping with the blueprint. In this manner, nanobots may be able to replace scar tissue with healthy heart tissue following a heart attack, and could repair crosslinked proteins, addressing one of de Grey's causes of aging. Although Drexler doesn't claim that nanobots will make us immortal, he does think they will help us live a very long time, in a physical condition of health and vitality.

Nanomedicine is based on the theory that if all molecular components, including the atoms, are arranged correctly, healthy function will follow. It's possible that this theory could be disproven with further research, but so far research has not negated it. The first nanotreatments will be the targeted destruction of cancer cells, followed by therapies aimed at ameliorating specific diseases, and then, if all goes well, we will enter the new paradigm of maintaining health and youth for a very long time, possibly hundreds of years. Some observers say that scientists won't be able to do the requisite research to make all this possible in less than a century. However, Drexler, Kurzweil, and others point out that soon, artificial intelligence, or the super-powerful computers just around the corner, will be able to fast-forward research many times faster than what human scientists can laboriously accomplish. For example, Kurzweil says, "It took us 15 years to sequence HIV; we sequenced SARS in 31 days," with the help of computers.[18] In the near future, machines will design and build nanobots far more quickly and efficiently than human scientists can.

Although many people are skeptical about radical life extension, we're already deep into a life-extending revolution. It's called modern medicine. Vaccinations, antibiotics, cancer treatments, bypass surgery—these all extend lives far beyond what a century ago was considered a "natural" life span. Clearly, the very concept of a natural life span is a moving target that has so far shifted rather slowly, but over time it has drifted dramatically. During the Roman empire, life expectancy was approximately twenty-five years, and one who lived to fifty was considered old. Today, we regard a death at twenty-five to be a tragedy, shocking in its untimeliness. So far, our extending life span has not been disruptive to society. On the contrary, society has adapted rather well because the change didn't happen overnight and has been accompanied by advancing technologies that made a growing population sustainable. It's possible that the extension of our life span may continue to be incremental and nondisruptive, but it's also possible that, given the unprecedented power of new medical technologies, we could see radical jumps in longevity in our lifetime. Such jumps will require major realignments in society, including a redefinition of "old age." I'll examine some possible social adjustments later in this chapter, but first I'd like to explore a little-recognized cause of life and health extension dubbed by the Nobel Prize—winning economist Robert

W. Fogel and his colleague, economist Dora L. Costa, "techno-physio evolution."

In 1997, Fogel and Costa published their findings from an extremely original study of the longevity gains of the last three hundred years. For their paper, "A Theory of Technophysio Evolution, with Some Implications for Forecasting Population, Health Costs, and Pension Costs," Fogel and Costa applied economic theory and statistical analysis to a huge body of historical information, including wartime draft records and pension and census records, which they correlated with health, body size, and longevity records of the past three centuries. Their most important observation was that technological advances in food production, manufacturing, transportation, trade, communications, energy production, medicine, sanitation services, and leisure time services are responsible for both the doubling of the life span and a huge increase in economic productivity. In the introduction to their paper, Fogel and Costa write, "over the past 300 years human physiology has been undergoing profound environmentally induced changes made possible by numerous advances in technology. These changes, which we call technophysio evolution, increased body size by over 50%, and greatly improved robustness and capacity of vital organ systems."[19] They assert that technophysio evolution is still very much in progress and is produced by "a synergism between technological and physiological improvements that has produced a form of human evolution that is biological but not genetic, rapid, culturally transmitted, and not necessarily stable."

Fogel and Costa took into account even conditions in the womb and the health and nutritional status of expectant mothers in their analysis. They describe historical conditions of not only low sanitation and high levels of infectious diseases, but chronic nutritional deficiencies that led to stunted growth, more diseases in adult life, and low economic productivity. Because of these conditions, people in Europe and the United States were considerably smaller than their modern counterparts. For example, in 1790 the average English male in his thirties weighed only 61 kilograms, or 134 pounds, and the average weight of French males was only 50 kilograms, or 110 pounds. These people were quite short by modern standards and much more unhealthy. The poorest 20 percent of the population was so chronically malnourished that they did not have the strength to work, which explains why one fifth of the population were beggars.

It's a widespread misconception that modern elders are being kept alive in a state of more age-related illnesses than their forebears. In fact, the stunted, malnourished people of the eighteenth and nineteenth centuries suffered far more health problems, including age-related conditions such as heart disease, and had less resistance to fight off the contagious diseases they were subject to throughout life. Fogel and Costa write that "young adults born between 1822 and 1845 who survived the deadly infectious diseases of childhood and adolescence were not, as some have suggested, freer of degenerative diseases than persons of the same age today; rather they were more afflicted."[20] Among the elderly in the nineteenth century, heart disease was about three times more prevalent than it is among today's elders, musculoskeletal and respiratory diseases were 1.6 times as prevalent, and digestive diseases were almost five times as common. Fogel and Costa attribute the high incidence of disease in middle age and later life to the long reach of poor development during the fetal and childhood stages.

The effects of chronic malnutrition and poor living conditions created a vicious cycle that profoundly affected one's health over a lifetime. They observe that "the basic structure of most organs is laid down early, and it is reasonable to infer that poorly developed organs may break down earlier than well-developed ones." They note that stunted growth and malnutrition may be "associated with variations in the chemical composition of the tissues that make up these organs, in the quality of the electrical transmission across membranes, and in the functioning of the endocrine system and other vital systems."[21] Many children suffered permanent neurological damage due to nutritional deficiencies in the womb, during infancy, and throughout childhood. Malnutrition is associated with a host of other conditions, including impaired functioning of the endocrine system, heart arrhythmias, and degenerative joint disorders.

Fogel and Costa correlated the number of daily calories consumed with height, weight, and health status, estimating that the average adult needed 500 calories a day just to support digestion and many more to mount an immune response to a contagious disease. Even those who were able to work had only small amounts of energy to do so. However, due to better nutrition, warmer clothes and buildings, and other improvements introduced by technology, between 1790 and 1980 the life span doubled and the efficiency of the average British worker increased by about 53 percent, creating a virtuous cycle in both health and prosperity.

Fast-forward to today. People who live in industrialized nations and quite a few developing ones are considerably taller and heavier, have much better health, and are far more productive than their predecessors. No small contributor to increasing longevity is the lessening in early life of malnutrition, infectious diseases, and poor living conditions. Modern technology and living conditions combined with greater access to better medicine have only begun to pay dividends in longevity. Add to this health-conscious nutrition, less smoking, and regular exercise (due in large part to more leisure time among the working classes) and we are guaranteed that technophysio evolution is still in full swing.

Since Fogel and Costa published their paper in 1997, major strides have been made in conventional medicine, not to mention biotechnology, artificial organs and implants, regenerative medicine, drugs, and a wide array of life-extending technologies that will greatly speed the rate of technophysio evolution. But one lesson is clear from their findings, and that is that we shouldn't focus on the later stages of life at the expense of its beginning. It's profoundly important that we provide every infant and child with adequate nutrition and good living conditions (a lead-free environment, for example) if we want to stave off a heavy burden of diseases as our population ages.

Another take-home message is that humans long ago started to engineer and control their own evolution through increasing control of their environment. "Culturally transmitted" evolution is taking place much faster than biological evolution, and it's having a pronounced effect on longevity. This is a good thing, because biological evolution cares little for making us live longer. It's really only concerned with getting us to reproductive age and helping us live long enough to raise children. Those whose genes predispose them to long life have no special advantage when it comes to the business of having and raising children and hence have no special position in the gene pool. Extending the human life and health span appears to be up to us, and we are well on our way to a dramatically improved situation.

The most common refrain against life-extending technology seems to be that emerging science is somehow "playing God." While we're certainly on terra incognita as we venture forth on this adventure, it's hard to see why gene therapies to fight diseases and aging are playing God any more than vaccinations against contagious diseases that would have otherwise removed millions of people from the gene pool. Those who oppose letting humanity take charge of its own evolution must, if they are to be consistent, oppose the last three hundred years of techno-physio evolution.

I mentioned above my belief that current attitudes toward aging may be based on outmoded stereotypes. This is based on the widespread misconception that our aging population is growing old at the expense of more disability, that old age is a time of loneliness and unhappiness, and that cognitive decline is inevitable. In 2013, the Pew Research Center conducted a survey of Americans to determine their attitudes toward radical life extension, which they rather modestly defined as living to age 120 or older. Fifty-six percent of the respondents said that they would not embrace radical life extension if given the choice.[22] The report describes Americans' reaction to the prospect of radical life extension as both "ambivalent and skeptical." Only 27 percent thought that life spans of 120 years or more will happen by the year 2050, and almost two-thirds think that "longer life expectancies would strain our natural resources." However, those who responded to the survey were also uninformed about the science of life extension. Only 7 percent had heard or read a lot about the science that is likely to extend life, and about 54 percent had heard nothing about it. Thirty-eight percent said that they had heard a little about radical life extension. While most of the respondents would like to live a little beyond the current median life span (78.8 years), only 9 percent would choose to live past the age of one hundred. I strongly suspect that most people, if presented with the prospect of living past age one hundred in a state of health and vitality, would have a different opinion. As it stands, even our conceptions about what the experience of aging is like today are riddled with misconceptions.

As mentioned above, throughout human history, old age has been a time of sickness and disability—not for everyone, but for a great many people. Life in past ages was physically arduous for even the young and healthy; there was not much incentive to live long in a feeble and decrepit state. In addition, retirement pensions and the social safety net are modern inventions. In earlier times, the average person had to depend on the limited resources of relatives to stay alive past their productive years. There was no effective medicine for the diseases of old age and contagious diseases were rampant. It was quite common for older people to die of pneumonia and other infections that would be easily curable today, yet in the nineteenth century Americans referred to pneumonia as "the old man's friend" because it was likely to cut short the suffering brought on by all the other afflictions of old age. Century upon century of conventional wisdom supported the view that old age is a time of illness and disability made worse by the decline of one's cognitive abilities. This is still true for some older people, but the picture is rapidly changing. It's only in recent decades that scholars have started to systematically study aging, and older people today are far better off than what conventional wisdom would have predicted.

For the average person in past ages, life for both men and women was predicated on the physical exertions of hard labor. Today, technology and industrialization have so changed and enriched human life that older people not only are more plugged in to the world at large, but have many interests to pursue through the Internet and a burgeoning nonprofit sector always in search of volunteers. Those who have some physical limitations can still be very much engaged in life if they choose to be. But that's not the only reason to be more optimistic about aging. A recent large-scale study by David Cutler, a Harvard economist who studies issues in aging and health, showed that we are not only increasing our life span, we're increasing our health span as well.

Dr. Cutler and colleagues examined data on nearly ninety thousand Medicare recipients from 1991 to 2009 and found a clear upward trend in the health of older people. He also affirmed what gerontologists have been observing for the last few decades, a phenomenon they refer to as the "rectangularization of the life curve." "Effectively," says Cutler, "the period of time in which we're in poor health is being compressed into just before the end of life. So where we used to see people who are very, very sick for the final six or seven years of their life, that's now far less common. People are living to older ages and we are adding healthy years, not debilitated ones."[23]

Research has shown that the last few months of life are the most expensive, medically speaking. People are likely to spend more on their medical care in the last few months of life than in the entire rest of their lives combined, and this is mainly due to heroic, technology-intensive efforts to keep them alive for a few more days, weeks, or months when they are critically ill. If one needs nursing home care, the costs are particularly high. But thanks to better medicine, says Cutler, many conditions that used to be debilitating no longer are. He cites cardiovascular conditions as the most obvious example. Thanks to bypass surgery and cholesterol-lowering drugs, heart attacks are occurring less frequently, and when they do happen, people today rebound much better than they did in past years. Cutler also credits the fact that people are much better educated about health and how to prevent or postpone age-related conditions than they were in the past. While a very old person today may not live to see radical life extension, the odds are she will live independently, and in reasonable health, until the very end of life, a goal that most of us ardently hope for. Given the strides made by conventional medicine, we can take heart that as medicine undergoes the coming revolution, there is a good chance that radical life extension will add not just years, but many more healthy and vigorous years to life.

There is a stereotype of older people that may underlie many Americans' lack of interest in radical life extension—the idea that old people are lonely, depressed, and unhappy. However, there's a substantial body of research showing that older people are not only just as happy as their younger counterparts, they're actually *more* happy. This may be counterintuitive, but it's a consistent finding across a large body of psychological research. Older people gravitate to positive experiences and are better at letting go of the negative. They derive more satisfaction from their social relationships and are better at solving interpersonal conflicts. They regulate their emotions better than younger people and their moods are more stable—all of which tends to confirm that there is something to the idea of wisdom in old age.

The prominent psychologist Laura L. Carstensen has long been studying the mental health of older people and has developed a theory to explain this phenomenon, called socioemotional selectivity theory (SST), or a "life span theory of motivation." Like many other researchers, Carstensen believes that the more positive attitude of older people is a matter of motivation, the decision to direct their attention to more positive information and experiences. In short, older people have a bias to focus on the positive. Some researchers have suggested that this phenomenon is a sign of cognitive decline and a necessity to focus on more simple, positive things, but Carstensen has a different explanation, one that revolves around our perception of time horizons. She says, "SST is distinguished from other life-span theories in that its principal focus concerns the motivational consequences of perceived time horizons."[24] In other words, our mental habits, whether positive or negative, are bound up by where we see ourselves in terms of our own life span.

By studying the emotions of people of different ages as they reported them throughout the day, Carstensen found that positive emotions were felt as frequently among old people as among the young, but that older people had far fewer negative emotions. Older people, it seems, are better at not ruminating on negative things and moving on. In addition, Carstensen identified what may be the nucleus of why this is so—a shift over one's lifetime from focusing on the acquisition of knowledge to a focus on regulating emotion. This shift is highly dependent on one's estimation of his remaining years of life.

In youth, when people perceive their remaining years to be open-ended, they focus their attention on seeking out as much information as they can gain. Carstensen writes, "they attempt to expand their horizons, gain knowledge, and pursue new relationships. Information is gathered relentlessly."[25] Older people, who view their time horizon as shorter, choose to seek out emotional satisfaction. "They are more likely to invest in sure things," says Carstensen, and that means deepening their existing relationships and savoring life.

Some bioconservatives such as Leon Kass and Daniel Callahan place a great emphasis on this life-savoring stage of life, made sweeter by the perception that time is fleeting. They raise alarms about greater longevity and the receding importance of death, predicting that if life were greatly extended, it would be far less valued. But one has to ask the question, what if we had both the information-gathering mind-set of youth along with the emotional satisfaction focus of age, and we had this experience over many, many years of life? Might there be new stages of life that we currently can't conceive of because of our short life span?

The validity of the intuitions of Carstensen, Kass, and Callahan that people are profoundly affected by the prospect of an approaching end to their lives is undeniable. Carstensen writes, "In several studies, we showed that younger people display preferences similar to those of the old when their time horizons are shortened, and older people show preferences similar to those of the young when their time horizons are expanded. . . . When conditions create a sense of the fragility of life, younger as well as older people prefer to pursue emotionally meaningful experiences and goals."[26] This suggests that it is possible for people to enjoy both perspectives, perhaps simultaneously. Might extremely long lives enable those who have achieved emotional stability and wisdom to keep pursuing new knowledge? What might people thus oriented contribute to society, culture, politics, and education?

In 2013, a German research team led by Florian Schmiedek published the results of a study examining overall cognitive performance in work-like tasks among people aged twenty to thirty compared to people aged sixty-five to eighty. Their results smashed another stereotype—that of the unproductive older worker. Testing of over two hundred people revealed that older people are not only more emotionally stable, they have more consistent cognitive performance throughout the day and from day to day than the young. Dr. Schmiedek explained, "Further analyses indicate that the older adults' higher consistency is due to learned strategies to solve the task, a constantly high motivation level, as well as a balanced daily routine and a stable mood."[27]

Another German researcher, Axel Börsch-Supan at the Max Planck Institute, demonstrated through studies of workers in the car industry that serious errors are much more likely to be made by younger workers than by older ones. Börsch-Supan said, "On balance, older employees' productivity and reliability is higher than that of their younger colleagues."[28] All very good news, and yet another illustration of the senselessness of age discrimination in the workplace.

But what about the general cognitive decline of the older person? Everyone knows that a failing memory, most evident in the classic "senior moment," or momentary brain freeze when searching for a word, is aging's cruelest blow. Or is it? Numerous studies have demonstrated that older adults lose speed in information processing and are slower to recall information stored in their brains. Until recently, few researchers have paused to ask why this is so. However, a groundbreaking study conducted by a team of German linguistic researchers made headlines in 2014, and it turned conventional wisdom about cognitive decline on its head.

One of the great plagues of life for the older person is that awful, pregnant pause when trying to remember a person's name. Because of the social importance of addressing people by their names, this is one of the most annoying and embarrassing senior moments. By applying what psychologists call psychometrics (the measuring of psychological variations such as knowledge and intelligence) to word retrieval skills in both younger and older people, the linguistic team, led by Michael Ramscar, added a new dimension to interpreting a huge body of data. They say that "psychometrics tests do not take account of the statistical skew of human experience, or the way knowledge increases with experience. As a consequence, when these tests are used to compare age groups, they paint a misleading picture of cognitive development."[29] In other words, comparative testing of old people and young people hasn't taken into account the vastly larger information load that older adults carry around in their heads. When tests are designed to factor in the time needed to process word recall in a much larger vocabulary, the difference in the speed of recall virtually disappears. To put it into a nutshell, "it does not follow that someone who can recall 600 birthdays with 95% accuracy has a worse memory than someone who can recall just six with 99% accuracy."

Ramscar and his colleagues reached their conclusions by data mining huge computer databases and running computer simulations of how the information-processing burden on older people affects speed of recall. They acknowledge that conventional memory tests have shown that older people take longer to recall words but demonstrate that much larger vocabularies and a lifetime of experience and learning pose a challenge to a seventy-year-old that does not encumber a twenty-year-old. They assert that memory tests are generally skewed toward younger people because they typically use frequently used words (those best known by young people with limited vocabularies) and don't include the more rarely used and unusual words known by older people. When trying to recall a word, the older test-taker must search through a hugely increased mental "database" that includes synonyms, words with a similar meaning, experiential associations, words that sound like the word searched for, words with a similar spelling, types of words and concepts that are somehow related, and a more finely tuned sense of the nuances of words. In fact, older people actually out-test the young in some categories such as fine-grained meanings, and can have double the vocabulary or more than the average twenty-year-old.

Ramscar started out his study taking the notion of cognitive decline for granted. He was surprised when the data showed over and over again that "older adults' performance reflects increased knowledge, not cognitive decline." The research team, in the conclusion of their paper, writes, "At the outset, we noted that population aging is seen as a problem because of the fear that older adults will be a burden on society; what is more likely is that the myth of cognitive decline is leading to an absurd waste of human potential and human capital."

While much more research needs to be done in this area, Ramscar's paper seems to have reframed the issue of aging and cognitive decline. At the very least, future studies of cognitive performance in older adults must factor in the huge increase in information that the older person has accrued across his life span. In addition, for society to reap the potential rewards of having more very old people, it needs to reevaluate its values and place more emphasis on accumulated learning, knowledge, and the synergistic form of intelligence we call wisdom.

The delay of aging and increase in health span discussed so far in this chapter have involved only the gains we're seeing from the progression of conventional medicine. I have cited these advances to give hints of what we might expect in a future world of radically extended health span. These hints are extremely conservative compared to the revolutionary transformation of society that we can expect once converging technologies and antiaging medicine really kick in. The new society that we will be creating may bear little resemblance to even the rapidly changing society of the twentieth century, a time of unprecedented social, economic, and technological change. Society will have to make profound adjustments to the presence of many people who live for hundreds of years, dwarfing today's life expectancy of almost eighty years. Eighty might possibly become the new twenty, and family relationships might include many generations of the same family all living at the same time. We may end up knowing and loving many generations of our own progeny, and the young might have a large group of relatives to provide guidance and support.

On the other hand, as people necessarily embrace multiple

careers, attend schools and universities perhaps several times during a long life, and even potentially have several marriages, each of long duration, our focus might extend well beyond the nuclear families we created in our youth. It's possible that we may not even be able to know all of our living relatives, both biological and by marriage. In the 2014 article, "What Happens When We All Live to 100?" published in *The Atlantic,* Gregg Easterbrook suggests that if people have very long lives, they might revolve around long-nourished friendships more than the nuclear family. He observes: "But if health span extends, the nuclear family might be seen as less central. For most people, bearing and raising children would no longer be the all-consuming life event. After child-rearing, a phase of decades of friendships could await—potentially more fulfilling than the emotionally charged but fast-burning bonds of youth. A change such as this might have greater ramifications for society than changes in work schedules or health care economics."[30]

It's possible that the future might finally deliver the true liberation of women. Greatly extended life spans would mean that women, because they will long outlive their reproductive and child-rearing years, will attain far more positions of power in politics, business, culture, and philanthropy. If the birthrate goes down dramatically because of shifting demographics, children might come to be more highly valued. If that happens, the job of raising children might finally come to be appreciated as the crucial job it is. Women who delay or interrupt their careers to raise children might find that their life experience adds something uniquely valuable that they can bring to positions of power. Also, if science succeeds in greatly extending the childbearing years, women might have greater control over whether and when to have and raise children.

One of the darker predictions among those who oppose radical life extension is that a large population of very old people will overwhelm social safety nets such as Social Security and Medicare and place an unsustainable burden on younger workers who must pay the bill. However, because older people are steadily getting healthier, it's more likely that we will extend our working lives well beyond the age of sixty-five. Even today, contrary to popular belief, numerous studies throughout the world have shown that healthy life extension creates wealth not only for older individuals but for the nations where they live. This is no doubt related to what we're seeing in both animal and human research—the "compression of mortality" trend, which means that those who are living longer are also remaining healthier until they reach the very end of life, a phenomenon that could spell huge savings in health care expenditures.[31] Once again, all this is happening now; with the advent of converging technologies and radical life extension, the longevity dividend is likely to become much greater. While numerous critics of life extension warn that it would entail possibly catastrophic rises in health care costs, CTs will most likely offset the potential for any major increase. Other clues in contemporary society provide good reasons to anticipate a more peaceful and law-abiding society. Ohio State University political scientist John Mueller suggests that older people are less enthusiastic about war, and since the population is aging in most nations, it's possible that we can anticipate a more peaceful future. In addition, crime rates would almost certainly go down since the majority of crimes are committed by younger people. And Gregg Easterbrook has noted that we might finally be able to outgrow our shopping-crazy consumerist culture. He writes that "neurological studies of healthy aging people show that the parts of the brain associated with reward-seeking light up less as time goes on. Whether it's hot new fashions or

hot fudge sundaes, older people on the whole don't desire acquisitions as much as the young and the middle-aged do. Denounced
for generations by writers and clergy, wretched excess has re-
pelled all assaults. Longer life spans might at last be the counter-
weight to materialism."[32]

We would be wise, in trying to envision what the world will
look like with radical life extension, to keep in mind that even
very old people will look and feel young. They may continue to
pursue many of the same activities as people many years younger
than themselves. The May-December romance could go to a whole
new level; today a 23-year-old has little in common with a
68-year-old, but as people begin to live for centuries, we may
find that a 150-year-old can be very compatible with someone
who is 220. Stark social and cultural divisions between people of
significantly different ages may vanish, and age discrimination
is likely to lessen. Social groups will likely coalesce around interests, hobbies, and professions. These are the easy adjustments,
but the way science and technology are developing, there may
be far more radical adjustments in store for the human race.

Some futurists talk about the prospect of immortality, but
for the purposes of this discussion, I'm assuming that biological
immortality is not within our reach. Nature provides the most
solid evidence of what can be achieved in my view, and we know
of no biological creature that is immortal. Even though there are
very long-lived creatures that don't go through a period of
senescence, every creature we know of eventually dies. Some
observers who address the issues of radical life extension seem
to speak of radical life extension and immortality almost inter-
changeably. Some, like philosopher Todd May, suggest that if
people were immortal, any life project would lose its urgency
and people would lose all drive to accomplish anything. But even
if our time horizon were eternal, there would still be a drive to

create a better future, because we will be inhabiting it for a very long time. I suspect that the same is true with very long lives. However, other objections to radical life extension abound.

Philosopher John K. Davis writes about the "Malthusian objection" to radical life extension, which points to bad consequences such as severe overpopulation and shortages of natural resources should life extension become widely available. The argument is that, even if life extension were to prove to be a benefit to many people, it would on balance diminish the well-being of the whole human race.[33] Davis, in a 2005 article published in the *Journal of Medicine and Philosophy,* considers several possible scenarios when weighing the ethics of radical life extension in a society in which some people want it for themselves and some do not. Because life extension would inevitably affect planetary resources, one's right to it must be weighed against the good of the whole. One choice that Davis considers is what he calls "forced choice." In this scenario, one would be able to choose radical life extension only if he or she agreed not to bring any children into the world, thereby offsetting her claim to the world's resources. This makes rational sense, but enforcing it would entail a greater curtailment of individual freedom than democratic societies are likely to accept. It's very hard to envision any Western democracy imposing the kind of invasion into one's personal life that would be necessary to ensure that life extenders don't have children. It would be far more palatable if those who choose life extension also choose to forego reproduction, but the reality is that most people want to have children.

If we accept that the Malthusian objection is valid and a future world with very long-lived individuals also suffers greatly from severe overpopulation, scarce resources, and the possible social and political unrest these things engender, many people may decide that they don't want radical life extension. Only if life meets a certain standard of quality will people find life extension attractive. In other words, if quality of life in a Malthusian world falls below a certain standard, we can expect that fewer people will choose radical life extension. If science and technology provide acceptable solutions to overcrowding, such as the ability of humans to leave the earth and populate other planets, it would be much harder to pose objections to life extension. But radical life extension will likely become available long before humans achieve such abilities, so while we can be optimistic about the future, we must still make our decisions based on the current reality of a resource-limited planet.

Of course, any projection about the widespread impact of radical life extension must take into account the fact that some people are not likely to choose it if given a conscious choice, and that opting out should always be an available option. Religious people, for example, might very well reject it if they believe in a heavenly afterlife that would be better than several hundred more years on earth. It's also possible that some serious diseases will prove incurable and discourage sufferers from choosing life extension. And severe mental illnesses like depression might turn a very long life into a veritable prison sentence. Other people, having lost cherished loved ones to death, may also opt out. There are numerous reasons why a person might decide to forego radical life extension before we even get to a Malthusian world of diminished resources.

One concern we should guard against is the possibility that a social and medical culture will evolve that assumes everyone wants radical life extension. Slowly, over time, more and more treatments that greatly extend life could become standard practice, and it could become harder to find what we might call "finite care," which would become a kind of alternative medicine. Suppose that insurance companies decided not to pay for non-life-extending care because it would not prevent costly advanced disease states. It's well known that today people are often resuscitated against their will and given heroic treatments to prolong life even when it means great suffering. Could the same medical-cultural bias extend into the future so that life-extending treatments are rendered to critically ill people when they don't freely choose them? And what if we choose to accept a radically life-extending treatment, and then something happens that makes us sorely regret it? Don't people have a right to die if they believe it's a better alternative to continued life? This question is beyond the scope of this book, but since we don't currently have a consensus on it, we should be careful not to foreclose the ability of every person to answer it for himself.

The two other scenarios that Davis considers are a moratorium on radical life extension until society sorts out all the ethical issues and making it freely available to anyone who wants it. He assumes that the act of denying life extension to those who want it unfairly denies them of hundreds of years of life, so using mathematics to calculate potentially lost life-years in those who want life extension to the years lived by people who don't want it, he concludes that the greatest good, in terms of human life-years, could be achieved by making life extension available to anyone who wants it.

At this point, the ethical issues behind radical life extension have already received a considerable amount of attention among bioethicists, demographers, scientists, writers, and other thinkers, but they have yet to be laid out clearly for the general public. This is highly problematic considering that people are already making choices, mostly unconsciously, that will greatly extend their lives. It's possible—even probable—that the options of modern medicine will continue to quietly skew toward greater life extension one treatment at a time, leading to long-term effects that not everyone would choose if they were fully aware. This is already leading to greatly complicated end-of-life scenarios such as those discussed in the early chapters of this book. To add just one more example, imagine if a seventy-nine-year-old man, in the middle stage of Alzheimer's disease, is found to have advanced colon cancer. Which response is better—to subject him to surgery and chemotherapy at the cost of great suffering in what may be the last clearheaded months of his life, or to not treat the cancer and allow it to progress until it takes his life before he becomes severely disabled from Alzheimer's? Many of us alive today will face such predicaments either in our own deaths or in those of our loved ones. In the absence of cures for some of our most devastating diseases, more and more of us will be choosing what we will die from.

In addition, at some point (or points), most of us will need to make choices concerning life extension, and the consequences of those choices will become more and more open-ended. For those of us who manage to stay alive, in the next few decades medicine will offer us an increasing number of life-extending options. Now is the time when we should be asking, Is radical life extension a good thing for individuals, for society, and for the planet we live on? So far we have been answering in the affirmative. If this were not so, there would be no more vaccinations or bypass surgeries. But we're on the cusp of an age when the average healthy life span could reach into the hundreds of years, posing profound challenges to the allocation of resources, the world economy, and the social fabric.

In 2007, Cato Unbound, the Web site of the libertarian think tank the Cato Institute, posted a discussion titled, "Do We Need Death? The Consequences of Radical Life Extension," in which four thinkers, two who answer yes to the question and two who say no, lay out some of the common positions on life extension.

In the first article of the discussion, Aubrey de Grey states that about 90 percent of us are killed by the aging process, and that all arguments boil down to whether we should fight not death but aging. After all, radical life extension would be a cruel joke if the aging process can't be arrested. De Grey is also clear in his position that the ability to arrest aging is not only biologically possible, it's a moral imperative because it will inevitably end a tremendous amount of suffering. He says, "We are close enough today to defeating aging that serendipity does not define the time frame: the sooner and harder we try to do it, the sooner we'll succeed. Thus, our inaction today costs lives—lots of lives."[34]

Diana Schaub, a political scientist at Loyola College, answers de Grey in an essay titled, "Ageless Mortals." Schaub writes, "scientists and doctors are searching for cures for the diseases that afflict us in later life. All of this is welcome. Right now, the human life span is 122 years (maximum life span is set by the longest-lived individual of a species). You might say that we are endeavoring to make life expectancy approximate life span."[35] She later goes on to note disapprovingly that emerging sciences "open the prospect of re-engineering the human life span." As I read this, it occurs to me that 122 is also an engineered life span, since the one individual known to have lived that long no doubt benefitted along the way from the blessings of modern medicine. After all, in 1900 we could have just as easily set the acceptable human life span at a much lower number, and in the days of the Roman empire, to a lower number still. It's not clear why 122 is an acceptable life span and 25, 150, and 200 are not. In fact, we will not know what the true limit to the human life span is until medicine has exhausted all of its alternatives.

Schaub, without saying exactly why, says, "I can't shake the conviction that the achievement of a 1,000-year lifespan would produce a dystopia." Lacking answers, she nevertheless goes on to ask questions designed to conjure up pessimistic emotions. Her most cogent argument is, "Life takes its toll on the spirit as well as the body. What would the tally of disappointments, betrayals, and losses be over a millennium?" She doesn't ask what the tally of successes, triumphs, and moments of contentment might be, but goes on to ask, "What would it be like to experience the continued vitality of the body in conjunction with the aging of the spirit? Would it mean the best of both worlds: the vitality of youth with the wisdom of maturity? Or the worst of both worlds: the characteristic vices of age with the strength of will to impose them on others?"[36]

Ronald Bailey, in response to Schaub, asks in the next essay, what, exactly, are the "characteristic vices of age," and what does she mean by the "aging of the spirit"? Schaub doesn't try to answer these questions, but Bailey points out that "declining psychic energy correlates pretty well with declining physical energy." As a more optimistic proponent of radical life extension, he says, "The 21st century will provide an ever-increasing menu of life plans and choices. Surely exhausting the coming possibilities for intellectual, artistic, and even spiritual growth will take more than one standard lifetime."[37] I can't help but agree with Bailey in my personal aspirations, but nevertheless note that not everyone is interested in great intellectual, artistic, and spiritual growth. It's critical that we protect the rights of those who aren't interested in a very long life to opt out of life extension, something that most bioethicists would agree with but that few discuss.

Daniel Callahan answers Bailey with an argument that leans heavily on the assumption that evolution and nature clearly intended for us to live to our current life expectancy and no longer. Among his assertions, he claims that "there is no obvious correlation between length of life and satisfaction with life," but as noted above, research has demonstrated that this is untrue.

The longer people live, in general, the happier they are. Callahan goes on to say, "it is not clear to me why utopians who hope to rid us of aging are not in a fairy-tale land of rationalization also—not necessarily for thinking that radical life extension is possible, but for thinking it will be good for us as human beings. I see no good reason to believe it will be, and of course—being speculative—no evidence can be offered that it will be. It is an act of faith, pure faith."[38] Again, Callahan is ignoring a body of research showing that there are clear rewards in living longer, especially if one maintains good physical and cognitive health. Still, there *is* a certain amount of faith involved in optimistic predictions about life extension, but Callahan and others are also relying on faith in their pessimistic view of the future. What we are dealing with is an unknown future. This has always been the case, and always will be.

Bailey accepts the uncertainties inherent in any forward-looking project as an unavoidable part of life. He writes, "Humanity did not solve all of the problems caused by the introduction of farming, electricity, automobiles, antibiotics, sanitation, and computers in advance. We proceeded by trial and error and corrected problems as they arose."[39] This uncertainty may be seen as a frightening obstacle to some bioconservatives, but others see it as part of the excitement of ever-unfolding discovery. So yes, to embark on radical life extension would require a leap of faith in the future and in our capacity to solve problems as they arise. However, as Bailey points out, those of us who expect to be around for a very long time will have more of a vested interest in solving big problems like war, poverty, and environmental degradation.

Callahan concludes his essay by saying, "Nature knew what it was doing when it arranged, through natural selection, to have all of us get old and die. That is the price of species survival and vitality, and it has worked well. I don't think we humans can invent a better scenario, but we can surely do much harm in trying."[40] In fact, as noted earlier in this chapter, biologists believe that nature did not select for any particular postreproductive life span. Nature selected for our survival up to reproductive and child-rearing age, and has remained more or less indifferent to what happens to us afterward. In addition, Callahan ignores the fact that we are already well into the human life-extension project. Through modern medicine and technophysio evolution, we already die much later than we would in the wild. If we accept Callahan's argument that our proper life span depends on what nature decrees, how can we justify the medical interventions that allow us to live to eighty years old today? He and other bioconservatives seem to be arguing for an arbitrary number of years that one ought to live, and that number just happens to coincide with the current life expectancy. But this number most certainly was not selected by nature. It's a product of civilization. Now civilization promises to extend our life and health spans even more. It seems that Callahan's central argument is that he trusts nature more than civilization, but the last time nature was in charge, we lived in caves, hunted wild animals for subsistence, and died in our twenties. If we wanted to reject civilization, it's already too late. The ship of human life extension has long ago set sail.

Our adaptation to radical life extension is likely to be just the beginning of a long, long journey of continued adaptations. Thanks to converging technologies, future societies may be not only multiracial, but composed of a large array of life forms not currently imagined. Biological humans might coexist with genetically engineered humans, androids, and beings that inhabit the gray areas between human and machine. In a world where humans go off-planet to worlds being cultivated for the future growth of humanity, we might have beings engineered for adaptation to conditions far different from those on earth, such as those who can live for extended periods under water or in extremely cold environments. There may be good reason in the future to create human-animal chimeras, whether it is to splice advantageous genes into the human genome or to alter nonhuman animals. There could be centuries of continual adjustment as "personhood" is bestowed legally, morally, and socially to an ever-increasing variety of beings, or such emergences and their accompanying adjustments may be necessary in an open-ended future of continuous evolution.

As exciting as these possibilities sound, they could be extremely dangerous if human beings don't change the more belligerent side of their nature. Throughout history we have been given to commit all sorts of crimes against those whom we consider different, whether they are ethnically, culturally, or politically different from ourselves. Imagine a future in which radically different species of beings compose a vastly more varied society. In order for this to work, human beings will have to embrace a cultural revolution that proceeds apace of the biological and technological revolution we have already begun. If we continue to seek to resolve conflicts through violence, terrorism, and warfare, the future will indeed be dismal. We are called upon to undergo cultural changes that we have scarcely dreamed of, and this may entail enhancing our intelligence to a state far above where it is today. We must move ahead to a condition above where we currently find ourselves, and that condition is likely to be transhuman.

# 8

## The Age of Social Robots

It's 8:07 A.M., and Victor, whom we met in chapter one, slumps through the door to his apartment, looking tired and miserable. He has just gotten off his shift at the Department of Urban Hygiene. His eyes look watery and his nose is red from blowing it. The lights inside the apartment flood the room and Hilda, his personal robot assistant, greets him at the door, having already been notified by Victor's car that he was about to enter the apartment.

"Welcome home!" Hilda says with almost-human enthusiasm. She stands five and a half feet tall and has a very streamlined female humanoid shape. Hilda exhibits a range of human emotions, but Victor chose her because she still retained a somewhat manikin-like appearance. When he was buying her, the more humanoid models he looked at on the Internet struck him as rather creepy, and the more humanlike they were, the creepier, like talking corpses or zombies. In addition, Victor has friends who seem to have developed a rather unhealthy attachment to their robots, and he wanted to avoid that. Hilda was just human enough to blend in while still letting you know she was the electronic marvel that she was, and not a real person. Truth in advertising, Victor thought.

Victor dumps his coat on Hilda's arms without a word and she goes to hang it in the entranceway closet. Her movements are fluid, but she has a somewhat exaggerated step in case she encounters an obstacle that she needs to step over. "How was your day?" she asks as she hangs the coat.

"Night," Victor corrects her, "and it was miserable. How was yours?"

"I ran a diagnostic on all the home systems and found that systems are all running within acceptable parameters and network signals are strong. I cleaned the kitchen and compressed eighteen ounces of trash without any problems. However, Goldie seems to be—" Victor interrupts her with an explosive sneeze and Hilda hastens to grab two tissues and hand them to him. He grabs the tissues, sinks onto the couch, and blows his nose loudly. Then he hands the tissues back to Hilda so she can dispose of them hygienically.

"You seem to be having some type of respiratory distress," Hilda says.

"I'm dying," Victor mumbles.

"A scan of your vital signs indicates that you're suffering from infectious rhinitis," says Hilda, who rather mangles the "r" sound when speaking in contractions, much to Victor's annoyance. "The chances of you dying from infectious rhinitis are 0.0001."

"I know that, you bucket of bolts. I was speaking figuratively!"

"Your use of the words 'bucket of bolts' suggests that I may have malfunctioned. Would you like for me to contact the manufacturer's technical support team?"

"I'm in no mood for *that*." He goes into the bedroom and soon reemerges in only his underwear and T-shirt, a habit of his that played into his choice of a less realistic-looking robot. He didn't want some lifelike thing staring at him when he indulged in this little eccentricity. "It's blazing hot in here, Hilda. Turn down the heat."

"The heat has been turned down by three degrees. Your scan indicates that you have a temperature of 100.1. Perhaps that's why you feel hot. And by the way, what sort of mood are you in?"

"A miserable one. What do you expect? My head is like a bucket of water, and it feels like I've been hit in the head with a sledgehammer."

"A scan of your head reveals no blunt force trauma and your level of hydration is within normal parameters."

"Thanks a lot!"

"You're welcome. Would you like some chicken soup?"

"No, I need to see a doctor. I could have meningitis. Call and make me an appointment. Immediately!"

"Your head scan did not reveal inflammation of the meninges."

"I don't care. This headache is excruciating." Victor rubs his forehead.

"Your vital signs suggest that you have a viral infection involving the upper respiratory tract characterized by congestion of the mucous membrane, watery nasal discharge, and general malaise, with an expected duration of three to five days." Before Victor can sneeze again, Hilda has handed him two tissues.

"*Aaaaahchoo!* A cold? That's impossible. This is not just a cold. Call Dr. Pinkerton and get me the next available appointment. Tell the nurse I want a face-to-face, and it's urgent!"

"I've already forwarded your symptoms and vital signs to Dr. Pinkerton. The office advises a decongestant, an analgesic, plenty of liquids, and bed rest. I'll bring you the pills."

"A decongestant? What a joke! I could be having a cerebral hemorrhage!" He blows his nose.

"Your head scan did not reveal—"

"I don't care. Something is wrong. I need to talk to a human being." It occurs to Victor that had he spoken to Elaine in the same tone of voice, he would be spending the night at the Electro-Lodge. But he couldn't quite control himself. It was Hilda's very accommodation to his every whim that had come to irritate him. And why did she always have to be right—no, the word is *accurate*. Hilda was always accurate, but in his book, she wasn't *right*. It was maddening.

"Hello!" says a playful, high-pitched voice. Victor's small robotic dog, a furry, convincing Pomeranian, has sidled up to him for a bit of interaction. "Would you like to pet me?"

"Not now, Goldie. I'm not feeling well."

"*Arf! Arf!* Okay." Goldie looks disappointed and turns to go to her "bed," plugging herself into her charging station.

"Hilda, get me a nurse in Dr. Pinkerton's office. I told you I need to talk to a human."

The face of a nurse appears on Hilda's chest-level touch screen. "I've been here for the last ten minutes, Mr. Suarez. I've prescribed a decongestant, an analgesic, plenty of fluids, and bed rest, just as your home companion told you."

"What? Why didn't you say something?" He grabs a pillow from the couch and quickly throws it over his briefs. He holds it as though someone were trying to pry it away. "Whatever happened to a thing called privacy?" His face is flushed with anger and embarrassment.

"If you're not dramatically better in five days, have your home companion call back." The nurse sounds rather surly. "And you don't have anything I haven't seen before, Mr. Suarez."

"Wait—this is more than a cold. Something is seriously wrong!"

"Your vital signs show a slightly elevated temperature,

which the analgesic will help. Goodbye." Hilda's touch screen reverts instantly to its home screen.

"Hold on—oh, hell!"

Goldie stands up and carries a ball to Victor, dropping it at his feet. *"Arf! Arf!* Throw the ball!"

"I told you not now, Goldie! Go to your bed!" Looking disappointed again, Goldie returns to her bed. Hilda hands Victor three pills and a glass of water, which she has retrieved from the kitchen. Victor knocks back the pills and sinks deeper into the cushions of the couch.

Robots like Hilda are at the epicenter of converging technologies, incorporating as they do artificial intelligence, wireless communication, engineering, linguistics, mathematics, imaging, neuroscience, and even psychology. Not only will such robots help to embed us even deeper with technology, but if mind uploading ever proves to be possible, some people will choose to have their minds housed in the bodies of humanoid robots. In the future, some of our primary relationships could be with robots. Hilda might seem futuristic, but all of the technologies used to generate her capabilities either exist or are in development. According to Bill Gates, numerous roboticists, and experts in artificial intelligence, we're on the verge of a personal robot revolution that can be compared only to the personal computer revolution of the 1980s.

Back in the '80s, many people asked, "Why would I need a personal computer?" The PC's many uses were not obvious until the technology became ubiquitous and the Internet matured. It would have been impossible for most people to envision the prominent role that e-mail and social media, for example, would come to play in their lives until it actually happened. Personal robots will probably follow a similar path as PCs and smartphones. They will enter many people's lives first through health care, but robots will also enter our lives through any number of well-established niches, including home care, self-care, communications, entertainment, therapy, and play.

Robots have long been commonplace in manufacturing, but lately there has been an explosion of innovation in the area of robotic surgery. Already more than 1.5 million people worldwide have been operated on using the da Vinci surgical robot. Even a few years ago it was hard to imagine how a robot could play such a critical role in the field of surgery, but the da Vinci system, which works as the surgeon's eyes and hands, amply demonstrates that machines continue to surpass humans in some important functions.

To perform surgeries using the da Vinci system, the surgeon is not standing over the patient. He's seated at a console, viewing a three-dimensional, high-resolution image of the surgery through a tiny camera while manipulating instrumentation that controls the steady, extremely precise movements of four robotic arms. The da Vinci robot, hovering over the patient, could look pretty menacing if the patient were awake, but these robots actually have greater dexterity with tiny, precision instruments than a human hand ever could, making for better outcomes than conventional surgery.

The advantages of robotic surgery are manifold. The miniaturized instruments eradicate potential hand tremors and make the surgery more precise, with less collateral damage. The use of miniaturized instruments means that incisions can be smaller than even those used in laparoscopic surgeries. To showcase the precision of the da Vinci robot, the da Vinci Web site features a video of the da Vinci peeling a single grape. In the video, tiny, delicate instruments carefully peel the skin away without any damage to the pulp—something that would be very difficult to do by hand. And the results of robotic surgery actually surpass those of either open-incision or standard laparoscopic surgery. Robotic surger-ies entail shorter hospital stays, faster recovery times, and fewer complications, which offsets any increase of the cost of equipment. The da Vinci system has now performed cardiac bypass, colorec-tal, gynecological, head and neck, lung, kidney, urological, and many other types of surgery with great success.

The makers of da Vinci say that robots will not replace surgeons anytime soon. Over time, however, more and more med-ical procedures are likely to be performed by robots. Since 2008, an anesthesia robot nicknamed McSleepy has been administer-ing surgical knockout drugs, and beginning in 2011, a remotely controlled robotic arm named the Kepler Intubation System (KIS) began inserting endotracheal tubes (breathing tubes used during surgery) into patients. Although the thought of having a robotic arm shoving a tube down one's throat sounds scary, KIS's performance has been shown to be safer and more pre-cise than when the same procedure is performed by an anes-thesiologist.

Surgery is just the tip of the iceberg when it comes to the ways that robots will be used throughout health care. Robotic prostheses are rapidly coming to the aid of people with limb loss in the United States, and robotic exoskeletons can assist in phys-ical therapy for the paralyzed. Delivery robots and search-and-rescue robots are able to enter dangerous situations that humans can't safely reach—rocky terrain, an avalanche, or a radioactive zone, for example. Disinfection robots will soon be able to sani-tize and disinfect hospitals and other areas where people can contract deadly infections, cutting down on an all-too-common cause of death, and pharmacy robots will soon be filling pre-scriptions.[1]

Robots could prove themselves very valuable in making medical diagnoses. IBM's computer Watson, which famously beat several human contestants on *Jeopardy*, is being tested for its diagnostic abilities at Memorial Sloan Kettering hospital. Although Watson doesn't have much of a bedside manner, the supersmart computer can process up to sixty million pages of medical text, including case studies, per second. He also has the ability to learn over time so that, theoretically, he will be able to draw from his own "experience" in diagnosing and making treatment recommendations. Although no one has yet made a systematic comparison, doctors sometimes make costly diagnostic errors. Presumably Watson will be better at diagnostics because of the enormous amount of medical data he can review. Watson-like robots may become indispensable assistants in every doctor's office, protecting both patients and doctors from the consequences of misdiagnosis.[2]

Of course, even though Watson can review and analyze a massive amount of information that real doctors couldn't possibly keep up with, Watson's trove of data could lead to many more tests than a doctor would normally order, driving up the overall cost of health care. As Jonathan Cohn observed in a 2013 article for *The Atlantic*, "If Watson tells physicians only what they already know, or if they end up ordering many more tests for no good reason, Watson could turn out to be more hindrance than help."[3] But Cohn also recognizes that health care, which already constitutes one sixth of the U.S. gross domestic product, is in for a quantum shift. Over seventy-seven million Americans are expected to retire in the next thirty years, and the aging of the population is already putting increased economic pressure on the health care field. One way we may be able to offset the costs due to a leap in medical demands is through the use of robots and computers like Watson. These computers could automatically and continuously update electronic medical records and analyze them in a way that busy doctors often don't have the time to do.

Many of the manufacturers of these robots assert that they will empower, not replace, medical workers, allowing them to focus on human interaction, but it's hard to see how jobs will not be eliminated. As usual, technological innovation will come with costs, and there doesn't seem to be a clear path to replacing a large number of medical jobs. Another issue is the substitution of the human interaction in routine medical procedures with robot interaction, perhaps leading to greater social isolation among the elderly and the handicapped. This problem is likely to be the most pronounced in what is perhaps the most important role of all—that of day-to-day personal care.

The first uses of personal robot assistants are likely to be as home health aides and daily living assistants for the chronically ill and the elderly. The robots will be able to take vital signs, remind people to take their medications, lift people from their beds, and perform other health-related functions, but they will also be able to provide critical services like housecleaning and doing laundry. While the vast majority of older people want to stay in their homes or "age in place," it is often their inability to maintain a household or help themselves with personal tasks like bathing, eating, and getting dressed that lands them in a nursing home. In the near future, robots will be taking over some of these tasks, allowing many people to remain in their homes longer, with more of their dignity intact than would be the case in a nursing home.

Once personal assistant robots enter the home, given the advanced state of current research, they will quickly prove themselves indispensable in a wide array of functions. In fact, Bill Gates predicted in 2007 that soon there will be a robot in every home. Writing for *Scientific American,* Gates noted that the robotics industry "is developing in much the same way that the computer business did 30 years ago. Think of the manufacturing robots currently used on automobile assembly lines as the equivalent of yesterday's mainframes." Gates observed that "We may be on the verge of a new era, when the PC will get up off the desktop and allow us to see, hear, touch and manipulate objects in places where we are not physically present."[4] Theoretically, the surgical procedures discussed above could even be performed by surgeons thousands of miles away from the patient.

The home health care robot is poised to take all of us into uncharted territory where robot-human interaction is as commonplace as our interactions with our smartphones. Robots are not only at the epicenter of CTs, they will soon become a huge part of the "Internet of things," the highly networked, interactive relationship of humans with their phones, computers, cars, wearables, and environment. Our personal care robots will be able to interact with all of the other technologies we use, making life safer and more pleasant and convenient. If they do become common household appliances, hundreds of millions of them will need to be designed, programmed, built, and serviced. The scale of the manufacture of robots could come to rival that of PCs and cars, and this could create a large number of jobs for some period of time. However, given the established trajectory of robotic technologies, it's likely that in time robots will even take over the manufacture of their robot brothers.

Robots promise to be life changing, but new technologies only become truly transformative when they become a cultural phenomenon—part of the everyday world of natural activities like communicating, working, playing, and being entertained. When this happens, robots will cross the line from industrial machines to captivating products in the consumer market. Humanoid robots will need to be emotionally assimilated with humans in order to be accepted in the highly personal roles they will play in the home, the office, and the health care setting. They will need to be customizable, learn from interacting with us, and develop relationships with us. All of this is being achieved through software algorithms that are enabling robots to learn through observation and interaction, the same way that babies and children learn. More on this later, but first let's take a look at how the robot industry is receiving a huge infusion of cash and muscle from the technological giants Apple and Google.

In December 2013, Apple announced its intended $10.5 billion investment in supply chain robots and automation equipment and confirmed that it had acquired the technology company Prime Sense for $350 million. Prime Sense makes 3-D sensing technology, a major component in robots. But Apple also seems determined to penetrate the personal robot consumer market as it did with its iPhones, iPads, and computers. It provided an important hint on just how it plans to do so by showcasing the toy company Anki during the keynote session at its Worldwide developers Conference in June 2013. So what is Apple's interest in a toy company, and why did it feature the company at such an important venue during its conference? The keynote event was all about Apple, as expected, until it came to a demonstration of four toy cars—yes, toy cars—racing around a track, powered by an app hosted on an Apple iPhone.

Four toy cars racing around an elliptical track is nothing unusual until you realize that the cars, rather than being controlled by remote control, are essentially driving themselves—sensing the environment in real time, adjusting to the shape of the track and to each other's positions, making "decisions" aimed at winning the race, and even trying to sabotage each other, driven by the software hosted on an iPhone app. At one point, three of the cars band together to try to block the fastest car by aggregating on the track in front of it. One by one, the fastest car blows each one of them off the track using one of its "weapons." Video gamers will recognize all of this from any number of racing games. What's different about Anki is that it has taken the video game off the screen and made it real. And it has taken small robots and tailored them to a commonplace human activity—play—and all using a handful of cheap components, none of which costs more than $1.20.

Following the demonstration of its robot cars at the Apple meeting, Anki's CEO Boris Sofman sat down with Alexis Madrigal, who interviewed him for a piece published in *The Atlantic*. During this interview, it became more and more apparent why Apple has taken such an interest in the small robotics company. Madrigal says, "For Sofman, entertainment, toys, are merely the way to get robotics into consumers' lives. He argues that their product is doing a lot of the same fundamental things that autonomous vehicles and other types of near-future consumer robots will do. And that they're merely taking the bottom-up approach to building out these futuristic capabilities."[5]

In fact, Anki's toy cars and its choice of entertainment as their entry point to personal consumption might easily be seen as a prototype for one way robots will be integrated into our lives. The toy cars are themselves examples of the kinds of converging technologies that are merging to create far more complex robots. "When you look at what goes into Anki Drive, what goes into this is: industrial design, mechanical engineering, electrical engineering, embedded systems, low-level firmware development, control algorithms, dealing with sensors, wireless communications, core robotics, artificial intelligence, mobile development," says Sofman. "Just getting the product together is a huge chain." By the "bottom-up" approach, Sofman means that Anki has merely taken already available technologies and components and put them together to do all of the basic things all robots will need to do going forward. "The core problems in robotics," says

Sofman, "positioning, knowing where you are, reasoning, using that information to make intelligent decisions, planning, searching, deciding what you need to do, and the execution where you need to move precisely in the real world—those carry over into any application in the real world."[6] Even as robots become more complex, able to do more things, they will all incorporate the essential abilities of positioning, making decisions, and executing those decisions.

Another aspect of Anki's approach—pricing—could not have been lost on Apple. Anki has built its bedrock on low-cost components—parts whose mass production has brought cost down to almost negligible significance. Sofman says of its cars, "There is no component in here that costs more than $1.20. We have cheap motors. A battery. A microcontroller, a 50 Mhz computer, and an optical sensor. Ironically [that sensor] is the front facing camera of an iPhone."[7] What's unique to the toys is the software that enables the core abilities of the game.

The real defining element of Anki Drive, however, represents a leap in functionality that promises to spawn a whole industry of new intelligent products that cut across categories like communication, entertainment, utility, and assistance—common activities in the real world. Companies like Anki are taking what formerly existed only in the digital or gaming world and relocating it in the real world, where real-life interaction can take place. Sofman seems to have his finger on the pulse of why robots will be so seductive. "There's a built-in desire for people to connect with things they can touch. It's more social. It's not as natural to look at something on a screen. You'll never replicate the connection you can make with something you can touch."[8]

Some will see a brilliant marketing sensibility behind Anki's approach while others may see something more ominous—a new class of technology cleverly insinuating itself into possibly every nook and cranny of our lives. This technology will sell itself through fun, entertainment, and convenience, and it will almost certainly offer us valuable services. It may enrich our lives a great deal and, through control and interaction, actually extend our own abilities in ways we have scarcely begun to imagine. But it will also become easy to become dependent on robots, perhaps too dependent. And interactive robots will take the blurring of humanity and technology to yet another level. Our social circles, and even what we consider to be our families, are destined to include not only cyborgs, but socially interactive robots.

Of all the companies investing in robotics in a big way, Google has by far made the biggest gamble that robots will soon inundate our lives. After hiring computer scientists and intellectual heavy hitters Ray Kurzweil and Peter Norvig as its director of engineering and director of research, respectively, in 2013 it acquired eight robotics companies, all of which are drivers of key technologies. Even more than most technology companies, Google is known for holding its cards close to its chest, but a few hints on where it's going can be gleaned from the nature of the companies it acquired.

The best known is Boston Dynamics (BD), which builds a wide array of robots, from all-terrain, wall-climbing, four-legged robots to humanoid-shaped robots. BD is well known for its DARPA-funded projects and projects for the U.S. Army such as Sand Flea, an eleven-pound, four-wheeled robot that can jump thirty feet into the air to overcome objects. Its best known robot is the humanoid Atlas, which was modified and used by several robotics companies that competed in the 2015 DARPA Robotics Challenge, an international competition among robotics companies to create a bipedal robot that can negotiate a number of challenges in a simulated outdoor disaster area.

For the DARPA challenge, twenty-three robots from five

different countries competed in tasks such as driving a Polaris all-terrain vehicle, disembarking from the vehicle, manipulating a doorknob, turning off a valve, climbing a wall, pulling a plug out of one socket and plugging it into another, negotiating a pile of rubble, climbing a flight of stairs, clearing debris, sawing through plywood, and using a fire hose. The robots were timed in how long they took to complete each task as well as whether they completed each one. DRC-Hubo, a sleek, versatile robot made by a team at the Korean Institute for Science and Technology, won the competition by completing all the tasks in just over forty-four minutes. But in the challenge, one major robotics company was conspicuously absent: Google.

Google's lack of participation was even more glaring considering that it had bought the Japanese robotics firm Schaft in 2013, just weeks before Schaft's bipedal robot won at a trial run of the DARPA Challenge. Despite being poised to win the real thing, Google took Schaft out of the competition, saying that it would not make its robots available for military purposes. It goes without saying that DARPA was looking for robotics suppliers outside of its usual supply chain, but some observers characterized the act as a snub toward the Pentagon and its industry rivals. Members of the robotics community attributed Google's actions to its well-known secrecy and a reluctance to share information with its competitors.[9] Whatever its reasons, the fact that robot design, programming, and manufacture are being advanced largely by the private sector means that competitive pressures will come into play. The closer any technology is to reaching the consumer market, the less information is likely to be shared.

Google's future plans for its robotics division might be extrapolated from the fact that three of its acquisitions specialize in humanoid robots. One is the aforementioned Schaft and another is Meka Robotics, whose robots feature humanoid heads rather than the stack of hardware many robots have in place of heads. Meka also makes humanoid hands, arms, grippers, and manipulators, and its robots seem promising as potentially interactive personal robots. While designing hardware that's appealing to humans is critical if robots are to enter the consumer space, the software is what really distinguishes one robot from another. *Robotics Business Review* columnist Celeste LeCompte, writing about Google's acquisitions in late 2013, notes that "the major link between the firms is that they are all software-intensive robotics companies." Writing about the exploding "Internet of things" trend, she says, "Robotics sits squarely in the middle of these trends, playing a critical role in blending the digital and physical worlds."[10] As our bodies come to harbor more and more technology, our personal robots will embed us even deeper into the networked technological world around us. The digital world is starting to step out into the physical world to interact with us in real time, a phenomenon that may make it harder than ever to distinguish between what is real and what's not.

I have devoted this space to a review of the recent activities of Apple and Google because they suggest that the two tech giants are poised to get into the humanoid robot business in a big way. Clearly their leaders see the enormous potential, both technologically and economically, of the personal humanoid robot as the next big consumer item. These companies have the resources to very quickly launch into the development, manufacture, and distribution of personal robot assistants for the home, health care, and entertainment. Much, however, is riding on their ability to develop the interactive aspects of robots that will allow them to be socially assimilated with humans.

For personal robots to be accepted into our lives, designers must master the psychology of natural human interaction, something that roboticists are already working on. A common term these days is "social robots," and there has been a considerable amount of research into what makes a robot socially appealing, something that will be of paramount importance in the case of personal assistance robots.

Research has shown that people, including the elderly, are surprisingly willing to accept robots as personal assistants. And one finding that has occurred time and again is that people are quick to attribute human emotions to robots and to assess them in the same ways they size up other people. A robot with an intelligent face is preferred in certain functions, such as giving financial advice, and a robot with a cute, childlike face is seen as being amusing and entertaining. In categories like personal health care, humanlike faces are preferred. However, there seems to be a limit on just how human people want their robots to look. Like Victor, many of us find a too-human-looking robot face strange and creepy.

Emotional intelligence is something we generally attribute only to humans, but the ability to simulate emotional intelligence seems to be a key characteristic in robots if they are ever to be accepted as personal and health care assistants. But even before designing a robot's social qualities, roboticists need to focus on the right kind of face for the specific role the robot is expected to play. As mentioned above, humans tend to attribute personality characteristics to the appearance of the robot's face, just as they do with humans. Akanksha Prakash, a former graduate student in psychology at the Georgia Institute of Technology, along with others, has already conducted a considerable amount of research into people's preferences for what sort of face they would like in robots of varying abilities. Prakash and a team of Georgia Tech researchers attributed their inspiration for a significant study called "Understanding Robot Acceptance" to the robotics company Willow Garage, a collaborator with Meka, one of Google's 2013 acquisitions.

In the study led by Prakash, the researchers found that older and younger people have different preferences for what they would like a personal robot to look like, and those preferences vary based on what the robot is expected to do. After examining the attitudes of both college-aged adults and older adults, Prakash found that both groups attributed certain social and personality characteristics to robots based on their faces. Most older adults preferred for robots to have humanoid faces while most of the younger adults preferred robots with either a robotic face or a mixed human-robotic face. But there was a clear demarcation related to what tasks the robots were expected to perform.

Both groups were pretty flexible in their preferences for robots meant to perform household chores such as mopping the floor or unloading a dishwasher. But for "intelligent" functions like helping to manage money, both groups preferred a humanoid face or a mixed human-robot face, which they associated with intelligence. The greatest differences between the age groups related to robots providing intimate personal care such as bathing. While humanlike traits were generally preferred when robots provided personal care, a significant number of the study's subjects "didn't want anything looking like a human to bathe them due to the private nature of the task."[11]

One thing that is clear from research is that older adults make the connection between robotic assistance and aging in place. Overall, they are highly receptive to the idea of having a robot to assist them with housecleaning and other home-maintenance activities. In fact, older adults are more accepting of robotic assistance than are health care workers and their own families. In another study conducted by the Georgia Tech team, older adults were embracing of robots to perform tasks like cleaning the kitchen, sweeping and mopping floors, making beds, and cleaning windows, as well as fetching hard-to-reach items.[12] Robot acceptance is clearly tied to specific functions and how they can impact a person's ability to live independently. Because of different preferences among different age groups and even between the sexes (women, in general, are more accepting of robot assistants), personal robots will need to be customizable and designed with a particular target market in mind.

The Japanese, who lead the world in robot technology, are exploring the role of companion robots for their rapidly aging population. The Japanese inventor Takanori Shibata has created a cute, furry, robotic seal called Paro and tested it out in nursing homes as a therapy animal. Paro, who coos happily when petted and squeals when handled roughly, was found to alleviate feelings of isolation among nursing home residents in much the same way that cats and dogs do. Paro also learns how to recognize his owner's voice and gestures from interacting with them.[13] It seems doubtful that the benefits of having a robotic pet can live up to those of having a relationship with a real animal, but these pets are undoubtedly cleaner and easier for the elderly to take care of than a flesh-and-blood animal. Paro has been tested out not just in Japan but in Italy, Sweden, and the United States, where he has demonstrated clinical benefits. So far, robotic pets are fairly simple, with a limited range of expression, but advanced algorithms that allow robots to learn the way children learn will soon allow them to develop a unique relationship with their owners.

The Japanese electronics giant Panasonic, formerly known as Matsushita, makes talking robots that look like cats or teddy bears. Empowered with voice-recognition software, the bots talk back when spoken to by elderly residents at nursing homes, even alerting a nurse if a resident doesn't answer when addressed. And Japanese baby boomers who want to monitor an elderly parent can buy a robot named Wakamaru. This talking robot with camera eyes is three feet tall and is programmed to be the eyes and ears of family members who can't be personally present, and it can do things like remind its charge when it's time to take her medicine. A similar robot can provide video face time so that family members can see each other and carry on a conversation.[14]

While the usefulness of monitoring a frail older person is clear, it's not so clear that older people will appreciate being watched twenty-four hours a day, even when the remote watcher is a loved one. There is something intrinsically unsettling about others being able to sign in at any time and use these devices to see and hear one's every move. Even though people are not being watched by a powerful entity like the government, many would feel that the right to privacy is an important asset that they're not willing to forego. And not all relatives can be relied upon not to use the bots in a malevolent way. If these bots become widely accepted for the care of the elderly, it seems likely that they will also come to be used to monitor children and teenagers. It's then only a short step to employers using the technology to watch workers and for law enforcement to watch citizens. Who will protect a citizen's right to not be watched? Will privacy become a thing of the past, or will societies develop safeguards to protect it?

Robots are destined to become the most important component in the highly networked "smart homes" that will be available in a few years' time. The Dutch company Smart Homes oversaw the multination development of a robot companion that was designed to be integrated with a smart home and smart clothes that can be worn by anyone for health monitoring. The entire system, called Mobiserv, includes a customizable, armless robot with a prominent touch screen, cameras, sensors, and audio capabilities for the cognitive support of older people. The bot reminds users to take their medicine or attend to other important things like eating or calling friends to stay socially stimulated. But the robot attains its maximum efficiency when networked with a home environment that incorporates things like smart sensors and optical recognition units, even bedsheets with sensors that can monitor a user's sleep patterns.[15] Robots like the Mobiserv bot will soon be able to turn lights on and off, change the home's temperature, turn TVs and speakers on and off, start dishwashers and other appliances, and provide a wide range of functions in addition to providing care and companionship to older people.

Although the benefits of a system like Mobiserv are obvious, the potential downside of such technologies is clear as well. The first consideration is who will be able to afford it, but once the system comes to be manufactured on a large scale, the cost of the robots is predicted to fall to about $4,500 U.S.[16] The Dutch makers of the bot insist that they don't see the Mobiserv robot replacing health care workers, but as these models become more and more capable, it's hard to see how they won't replace at least some day-to-day caregivers. If government health agencies such as Medicare and Medicaid pay for them, as they might do, considering the cost savings of relying less on human caregivers, more caregivers will be put out of business. Another issue might be the programming and maintenance of the robot and the likelihood that users will lose competence in managing their smart homes through dependence on the robot. Should the robot malfunction, it's possible that other components of the smart home will be affected, and the user might be helpless until a technician comes to remedy the problem. And there is also the possibility that glitches and the complexity of using such a system could make the older person's life more—not less—complicated. And of course, the Mobiserv will have the ability, welcome or not, to monitor users twenty-four hours a day.

No matter how you look at it, the Mobiserv bot is replacing functions that have traditionally been provided by real people and that form the warp and woof of many important relationships. Older or sick people will almost certainly become highly dependent on them, and busy family members might over-rely on them to provide companionship to their loved one. No matter how friendly the robot is programmed to be, the user will not be in a relationship with a real person, though the boundaries between real and unreal will surely blur. The quality of the relationship can hardly be compared to that of one with a real person, since the endless compliance of the robot will not socially challenge the user. Human relationships tend to be messy and unpredictable, but it's often these very qualities that make us grow. Would we choose authentic human relationships when our relationship with a robot is far easier and more convenient? In fact, the longer the relationship with the robot exists and the more one depends on it, the greater the risk that the user will lose important social skills such as unselfishness and respect for the rights of others.

The more advanced the personal robot becomes, the greater the danger that users will come to over-rely on them and to concomitantly lose social skills. With the rapid evolution of artificial intelligence will come robots that are ever more humanlike and difficult to distinguish from real people. Robots may come, for some, to replace even primary relationships like those of a parent or a spouse. Real people might be emotionally displaced when a spouse, for instance, starts to prefer a sexual relationship with a robot that caters to his every whim. Robots' abilities to learn from observation and trial and error will make them seem intelligent and authentic. They will adapt to the user's personality and the relationship will become highly customized. Robots, in effect, might become "better" companions than humans. As robots become more humanlike and humans potentially become more robot-like, the line between woman, man, and machine may become barely perceptible.

"Being a cyborg isn't just about the freedom to construct yourself. It's about networks," says philosopher Donna Haraway.[17] Humans will soon harbor implants and bionic parts that will interact seamlessly with their own Internet of things, including a highly abled, intelligent robot. In fact, through this intricate networking, we will each have powers and capabilities that far transcend our natural abilities. Human identity and personhood will be expanded to include a broad electronic ecosystem. While philosophers will continue to debate the issue of what composes a human being, robots are already performing in ways that urgently demand some definition of their legal rights and responsibilities, and this issue hinges on whether or not to designate them as persons. Robots already trade stocks, land airplanes, sell goods on Amazon.com and eBay, serve the military, and determine our eligibility for programs like Medicare and Medicaid. But the robot that is really pushing the issue of legal status (and ethical responsibility) is the self-driving car.

Google's self-driving car has already logged over 250,000 miles on U.S. roads, so far with few problems. But what if the car were involved in an accident that caused injuries to people and property? Who would be liable—the driver, who was not actually operating the vehicle? Google, which designed and programmed the car? Or even the car itself? From whom would the injured collect damages, and if traffic laws were violated, who would receive a ticket? The judicial system's tool for adjudicating responsibility depends on the recognition of personhood, and several legal experts are recommending that robots, including drones and autonomous cars, receive a legal designation of limited personhood. Even the aforementioned Watson might be designated as a person with legal rights and responsibilities should he recommend an incorrect diagnosis that leads to injury or death.

John Frank Weaver, attorney and author of the book *Robots Are People Too,* has outlined his recommendations for robots in *Slate*. He thinks that robots should be given a legal designation of limited personhood, like the limited rights and obligations of corporations. He lists five rights and obligations that should be granted to robots to protect humans from things like frivolous lawsuits and to give victims of damages a way to quickly recoup their losses.

The first right Weaver would grant to robots is the right to enter into and execute contracts. For example, having an Amazon drone deliver a product to your doorstep would entail entering into a small contract that indemnifies Amazon should the drone mangle the package or deliver the wrong thing. Since Amazon would have played no part in the mistake, it would have no legal liability. That would rest on the robot. A closely related responsibility for the robot is the obligation to carry insurance.

Weaver uses the self-driving car as an example of a robot that should be insured. If these cars carried insurance (purchased by the car's owner), the owner could not be sued if it causes an accident, and this would be an incentive for people to buy self-driving cars. The injured would be able to quickly receive an insurance payout while owners would be protected from frivolous lawsuits.

Looking down the road to a time when robots and other machines will be able to create art, music, literature, and the like,

Weaver foresees that robots will have the right to own intellectual property. He recommends limited intellectual property rights under which the inventor of the robot or program holds a patent on all its creations for ten years, after which the work will enter the public domain. This would incentivize inventors to design creative programs while not allowing permanent ownership of artworks that the machine may have autonomously created.

Robots should also have the obligation of liability in certain circumstances. This makes the carrying of insurance meaningful, and in the case of a car accident, it's the vehicle's insurance that pays out damages, not the owner's. The last recommendation anticipates a time when parents rely on robot nannies to watch their children. While the advisability of this practice could be the subject of another book, Weaver thinks it is bound to become commonplace. He lists among a robot's rights the right to become the guardian of a minor. Before we panic over the idea, it should be recognized that, for legal purposes, the custody of a child can pass through several entities in a single day. He says, "in our legal system, a child is always in someone's custody. Kids pass from their parents' custody to their school's custody, to the babysitter's custody, and back to their parents again every day."[18] Giving temporary guardianship to a robot nanny means that the robot will be liable should something go awry. In addition, making robots liable in this way encourages manufacturers to create the safest models possible.

Some people will no doubt object to the ethical implications of granting limited personhood to robots, but legal experts who are making this recommendation are thinking within the framework of already existing laws. Mark Goldfeder, a senior lecturer at Emory University School of Law, notes that "legal personality makes no claim about morality, sentience or vitality. To be a legal person is to have the capability of possessing legal rights and duties within a certain legal system, such as the right to enter into contracts, own property, sue and be sued. Not all legal persons have the same rights and obligations, and some entities are only considered 'persons' for some matters and not others."[19] The legal precedent for limited personhood has been firmly established in the case of corporations. Goldfeder doesn't think we should wait for robots to closely resemble us before we grant them the limited rights of personhood. "The establishment of personhood," he says, "is an assessment made to grant an entity rights and obligations, regardless of how it looks and whether it could pass for human."

Because a robot is granted limited personhood under the law, however, doesn't mean that for social and ethical purposes we would define them as "human." One of the clear advantages of robots is that they can operate in more dangerous conditions than humans could endure, such as in radioactive zones or in the more dangerous scenarios of war. So far, no one equates the loss of a robot with the death of a human being. However, as robot technology becomes ever more advanced, robots might accrue an increasing number of rights, including the right not to be deactivated. There may be a time when the deactivation of a robot will be considered a crime, and when the protections of robots become similar to those for people. Part of the determination of robot rights may well take into account consideration of the deeply enmeshed relationships of robots with human beings. Deactivation of a robot may become nearly as distressing as the death of a human loved one.

One issue that will certainly muddy the water, should mind uploading ever become a reality, is the status of uploaded minds, whether they are housed in robot bodies or not. Legally and ethically, will these facsimiles step into the same rights and responsibilities of the original person? Would the uploaded entity have inheritance rights, become the legal parent of children or the spouse of a surviving wife or husband, or step into the deceased person's job—or would it begin "life" with a blank slate?

Another scenario that could be troublesome is when a detailed facsimile of the mind of a living person is made while the original is still alive. The prospect is not totally out of the question based on the direction of current science. Researchers are now at work on the first virtual simulation of the brain beginning with a scan of a brain's neural structure and using such a scan to figure out how the brain works. Would legal personhood apply to digitized copies of human brains? Could a copy of a person exist simultaneously with that person, and what would its legal rights be? Would the facsimile be able to copy itself, or could the original person make multiple copies of himself? Would he "own" his copies, and what would he be able to do with them? These questions deserve a thoughtful examination that exceeds the purview of this book, but we should be considering them now.

What about the possibility of dangerous or destructive robots? The longtime staple of science fiction, the robot that turns on humanity, is something that some researchers are already trying to circumvent by programming robots with some type of moral framework. The idea of moral robots is not so farfetched when we consider the many uses of robots. Even the driverless car may encounter morally demanding decisions that even a human would be hard-pressed to make.

Imagine, for example, that a driverless car detects a young woman pushing a baby stroller on the side of the road, but if it were to swerve around her, it would crash into a car carrying two passengers. Does the car hit the mother and baby, or swerve to the left, hitting the other car? And what about a robot assisting in an evacuation in a situation of war or natural disaster? If there are multiple victims, how does the robot decide whom to save or treat first? Some type of ethical programming will need to be applied to robots in the near term, not the distant future.

Once again, the military is at the leading edge of technological innovation. The U.S. military is funding a collaborative effort among Tufts University, Rensselaer Polytechnic Institute, Brown, Yale, and Georgetown University to figure out how to imbue autonomous robots with an ethical sense. The project is a sign of the growing reliance on military drones. So far the drones have all been remotely controlled by human soldiers, but military research is trending toward autonomous weapons that will need to make lightning-quick decisions about who is a combatant and who is not and what collateral damage is acceptable.

Military robots will no doubt be the proving ground in determining if machines can make moral decisions, and the military has a head start in this endeavor. A great many military decisions are based on internationally recognized rules of engagement. Georgia Tech artificial intelligence expert Ronald Arkin, who wrote the book *Governing Lethal Behavior in Autonomous Robots,* thinks that military robots will actually make better decisions than human soldiers, precisely because they are not subject to human foibles like anger and a taste for revenge. In addition, robots will be able to quickly consider all the possible courses of action to pick the right one. One the other hand, robots can't be programmed to cover every possible scenario that could erupt in real life. Suicide bombers and other terrorists may be able to elude autonomous robots by blending in to a civilian population, which calls to mind a possible moral quandary. Suppose a wanted terrorist, dressed in civilian clothes, takes refuge in a hospital or a school. Should the robot strike the target in spite of the fact that many innocent lives will be lost? Will it be able to weigh the possibility of innocent lives spared from future terrorist actions against the certain loss of present lives?

Critics of the effort to design moral robots assert that, even with a huge amount of programming, situations of war or disaster simply present too many variables for a robot to effectively cope with. The robotics expert Noel Sharkey says that robots, no matter how sophisticated, will never have moral agency—the will and ability to make truly moral decisions.[20] It's not uncommon for even human beings to be unable to formulate a decision about the right course of action when a number of choices are possible. Part of the problem is that there are too many variables in the real world for humans to anticipate every possible outcome from a range of decisions. And those things that can't be anticipated can't be programmed into machines.

Machines may not ever be able to know right from wrong. They may only be able to choose one of many courses of action that don't incorporate unpredictable circumstances. But putting autonomous robots into life-or-death situations means they will be making life-or-death decisions. Roboticists need to collaborate with not only engineers, linguists, programmers, and psychologists, but also with ethicists, lawyers, and policy makers. One way to circumvent problematic results would be to not put robots into life-or-death situations or to keep all of the decision making in the hands of humans. But one of the chief reasons for using robots rather than humans in dangerous situations is that robots are considered expendable. With the further evolution of robotic technology, though, some robots may attain a higher value, especially when they have gained valuable experience, including intricate relationships with human beings. We might put these social or personal robots in a higher class than that of their more utilitarian brothers and treat them accordingly.

While scientists and philosophers grapple with the problem of making robots ethical, we humans will be grappling with the issue of our growing dependence on robots and other technologies.

Perhaps the most important skills people might lose through dependency on robots are our social skills, and this could inflict great social and psychological harm. Being cared for primarily by robots from the cradle to the grave means that critical family and social ties could be undermined. Will we still need each other when robots become our nannies, friends, servants, and lovers? These dependencies will probably be integrated into our lives quietly, with little fanfare, and by the time we start to realize the harm to our social, emotional, and spiritual development, it may be too late. We may already be too dependent on robots to extricate ourselves because we will have forgotten how to have human relationships. Today may be the last vantage point in history when we can evaluate the uses of robots without the blinders of dependency. Will robots make us more powerful, or will they be our downfall?

# 9

## Just Don't Call It Transhumanism

Today Transhumanists, with a capital *T,* eagerly embrace all of the technologies covered in this book, and then some. The inclusion of mind uploading and cryogenics (the freezing of deceased people for later thawing and reanimation) in their list of hoped-for technologies has placed them outside of the scientific mainstream, and they have in recent years been seen as a weird cult. But as Transhumanists have sought to lose the element of cultishness and be taken more seriously, the mainstream research world has also quietly moved closer to a transhumanist point of view. Several Transhumanist thinkers, such as the Oxford philosopher Nick Bostrom, hail from prestigious universities and write thoughtful, first-rate essays arguing for a human future of "boundless expansion, self-transformation, and dynamic optimism." With the advent of converging technologies, Transhumanism is becoming more respectable, and transhumanism, with a small *t,* is rapidly emerging through conventional, mainstream avenues. Like our character Victor, countless people who would recoil from the label "transhumanist" are accepting more and more of the therapies offered by converging technologies.

As revolutionary as it is, transhumanism isn't coming with

trumpets blaring—in fact, the average person has never even heard of it. But as Francis Fukuyama wrote in 2004, "Transhumanism of a sort is implicit in much of the research agenda of contemporary biomedicine."[1] Transhumanism is coming to us through doctors, hospitals, universities, and other research institutions, the U.S. military, and NIH-funded labs. Most of us do not consider ourselves Transhumanists with a capital *T*, but the technologies that fall within that category speak to an ancient quest for more life, strength, beauty, and ability that is well-nigh universal. Because transhumanism's entry point into our lives is currently through medical therapy, it's not likely that the trend will be reversed. Artificial organs, pacemakers, and other assistive devices and psychotropic drugs are already in use, and the embrace of transhumanism is largely unconscious. The conscious choices we're making are between suffering and relief, disability and ability, death and life. It just so happens that human enhancement and life extension come with the package.

Opposition to transhumanism also has deep roots, most prominently in traditions that place a taboo on hubris and on religious interpretations that see man as capable of abrogating the prerogatives that should belong only to God. Yet most, if not all, religious traditions place a premium on perfection attained through spiritual transformation, and conservative ethicists fail to reconcile the high value they place on perfection with the taboos against radical self-improvement that they favor. One thing that bothers bioconservatives is that the search for perfection is being sought not just in the spiritual hereafter, but in the earthly here and now. They sanction the search for perfection when it pertains to the spiritual life, but fail to prove that there is an inherent conflict between one's spiritual aspirations and seeking the best possible life on earth.

Human enhancement is welcomed by a virtually universal

psychological substrate that prods us toward continual improvement. In many people, this drive is opposed by a fear of the future and a dread of punishment should we tread across traditional boundaries. For some, fear will outweigh the impulse to seek out radical self-improvement, but for many others, the lure to become healthier, stronger, smarter, and longer-lived will prove irresistible. Having, to a great extent, mastered our environment, we now stand face-to-face with our own limitations, which stand between us and the perfection we crave. It can be argued that we will likely never achieve a state of perfection, but it can also be argued that there is something invaluable about the striving itself that has paramount meaning.

Fukuyama claims that the democratic concept of universal equality rests on the assumption of a human essence that transcends race, gender, and even intelligence, and that "modifying that essence is the core of the transhumanist project."[2] But Fukuyama and other bioconservatives have been making similar claims for decades without ever defining the human essence, and without that definition, their arguments remain hazy. Bioconservatives have gotten away with such arguments for a very long time because there has been no way to test their hypotheses. However, now that technology can make radical changes to the human organism, and is on its way to creating highly advanced artificial intelligence, many of our assumptions about human nature will soon be tested. If I can radically enhance my memory, decision-making capacity, and mood while staving off all tendencies to mental illness, what, then, is my essence? On the other hand, what if I make all those changes and find that my subjective consciousness has changed very little or not at all? In the last analysis, these questions can't be answered without cautiously moving forward with the human enhancement project and experiencing the results firsthand.

In explaining his objections to human enhancement technologies, Leon Kass has written extensively about what he calls "the wisdom of repugnance." He writes: "In crucial cases . . . repugnance is the emotional expression of deep wisdom, beyond reason's power to fully articulate . . . we intuit and feel, immediately and without argument, the violation of things we rightfully hold dear . . . generalized horror and revulsion are prima facie evidence of foulness and violation."[3]

Something about Kass's argument makes sense; there is such a thing as having strong feelings that are valid even though one can't quite articulate them. But it's the job of bioethicists to articulate feelings so that they can be examined in the clear light of day. For Kass to rest multiple assumptions on a feeling he can't explain, even upon reflection, does nothing to buttress his position. If one has a valid reaction of repulsion, that reaction should stand up under examination. In addition, those things that revolt one person may be perfectly fine to another. Some of the parishioners at Stacie Sumandig's church were repulsed by the sound of her artificial heart, but to her and her loved ones, the heart was a priceless gift. One can initially be repulsed by a new idea, then realize with further examination that there is nothing to oppose or be afraid of.

With his assertion of repugnance in general, Kass treads dangerously close to a defense of blind prejudice. We need to keep in mind that repugnance varies from culture to culture. To some cultures in the Middle East, the sight of a woman not covered from head to toe is repugnant, yet such a reaction is impossible to justify to the rest of the world. Many individuals and some entire societies are revolted by an endless list of new technologies, ideas, customs, and mores. It's possible that the conservatism that spurns anything new is part of the glue that holds society together. But if the notions engendered by that conservatism can't stand up to examination, the argument falls apart.

Some of those who oppose human enhancement don't just reject it for themselves; they believe that the dangers are so great that enhancement should be denied to everyone. But the desire to prevent people from adopting radical enhancement can also be seen as stifling humanity's natural impulses to grow, create, and make progress. While any reasonable person would agree that everyone has an inalienable right to reject enhancement for himself, it takes a stronger argument to persuade that he has a right to withhold enhancement from others who desire it.

There are other inconsistencies in the common arguments against radical new technologies that expand human ability. The theologian Ted Peters, in his 1997 book *Playing God? Genetic Determinism and Human Freedom,* examines an irrational tendency that is at play in many arguments opposing advanced technology. From Peters's point of view, humanity can't possibly steal power from God. Nevertheless, religious bioconservatives seem to assert that God is omnipotent, yet at the same time human beings can abrogate his power by taking their evolution into their own hands—a contradiction. There is also a revolt against any effort to make life on earth more attractive because of the unspoken assumption that this would somehow steal some of the glory from heaven. But if heaven is what it's cracked up to be, humans couldn't possibly build a world that competes with it. Such inconsistencies as the concept of an all-powerful God whom humans can still somehow diminish can't hold up to the light of examination. Despite the clear objections to transhumanism by bioethical conservatives, the world's religions so far show signs of developing a more nuanced point of view.

Although some see an inherent conflict between religion and human enhancement, there may be common ground that the two approaches to human life share. As mentioned above, many religions hold sacred the quest for human perfectibility. Followers are encouraged not only to seek perfection in spiritual transformation, but to live the best possible life on earth. This means primarily the best spiritual life, but many traditions teach that deep spiritual development leads to improvements in all aspects of life. In addition, since therapy and enhancement are so intertwined, common ground can also be found in the area of easing the suffering of the sick.

The world's major religions have barely begun to consider their positions on human enhancement, but many of them have started to contemplate radical life extension, which is an integral part of human enhancement. Their thoughts on life extension hint at how these traditions might establish a consistent view with enhancement in general. In 2013, the Pew Research Center canvassed thinkers from a wide range of religions and Christian denominations to find out how each might approach the issue of radical life extension. If there's a common thread, it's that the alleviation of illness and suffering is desirable, but that physical immortality is not because it circumvents rebirth on a spiritual plane.

Several Christian denominations put a high value on living a healthy lifestyle, connecting it to a better ability to further the church's mission. Allan Handysides, who is director of health ministries for the Seventh-Day Adventist Church, says, "The longer we live and the healthier we are, the better we can do our work."[4] Like virtually every other religious group canvassed by Pew, the Seventh-Day Adventists would be concerned about the equal distribution of life-extension technologies. This concern is well-nigh universal among religions, and if history is any indication, it won't stop the development of cutting-edge medical technologies. However, this concern could end up being short-lived. It's probable that like any new technology, life-extending and life-enhancing technologies will be available first to the rich and that there will be a certain lag time before everyone has access to them. It makes no sense to stifle innovation because of an anticipated delay to universal access.

Over time, human enhancement will probably become available to everyone, especially as long as it remains within the framework of medicine. The blurring of the lines between therapy and enhancement is actually a good thing for the time being, because it increases the chances that such treatments will obtain insurance coverage, whether private or governmental, and it primes the pump of public opinion to start regarding enhancing therapies as a right just like any other medical technology. All it would take is a widely accepted shift in the expected life span to medicalize earlier death and thereby make it amenable to covered treatment. In 1900, it's likely that no one considered death at sixty as unnatural, but today we would regard such a death as something untimely that must be prevented of at all possible.

Other major world religions, while not having an official position on life extension, are generally positive in their views. The Jewish ethicist and theologian Barry Freundel says, "Judaism has a very positive view of life . . . so the more of it, the better. . . . The goal in Judaism is to make the world better and [extended life] would allow us to do more of that."[5] Muslims believe that God has an inviolable plan for every person, says religion professor Aisha Musa. "Since you can't really violate God's plan for you," she says, "life extension is all right because it's part of God's will. . . . Whatever we do, God has a hand in it."[6] Muslims would likely not seek immortality, though, because "there is a deep-seated belief that death is a blessing."[7]

According to James Hughes, who is a former Buddhist monk, "Dramatically longer life would be beneficial, because it would give each person more time to learn wisdom and compassion and to achieve nirvana."[8] But a long life may be a double-edged sword, according to the Buddhist nun Karma Lekshe Tsomo, if one did not live a virtuous life. A longer nonvirtuous life would increase the accumulation of bad karma, whereas a longer virtuous life would enable one to work off bad karma while accumulating the good.

Whatever position we hold on the wisdom of human enhancement, we have already embarked on much of the transhumanist journey, and society has long ago accepted the basic premises underlying it. We have almost always accepted any enterprise to enhance ourselves, whether it's by education, training, meditation, drugs, or any other conventional means. We have accepted genetic manipulation through arranged marriages and selective breeding; we have embraced life extension through technophysio evolution and modern medicine; we have accepted artificial devices being integrated into our bodies through pacemakers, deep brain stimulation, implantable defibrillators, and the like, and we've accepted virtually all the techniques meant to ameliorate physical pain and psychological suffering. Transhumanism is already happening all around us, and we're welcoming it into our lives. But we have yet to sort out how we can adapt to a quantum leap in human life and ability in a way that's just, fair, reasonable, and sustainable.

It's already time to begin some of the major adjustments that transhumanism calls for. In order for converging technologies, which play such a huge role in transhumanism, to be truly liberating, society will have to address the idea that we all have a right to die when the further extension of life becomes too much of a burden. This is going to be an uphill struggle in and of itself; in American society, there is no more pervasive taboo than the discussion of death. And as discussed in the early chapters of this book, there is a well-established medical bias against "giving up" and letting a person die, even when further treatment does nothing but cause great suffering. However, not only must we accept each person's right to die, but we will have to grapple with decisions about how each person will die, and like it or not, we will have to learn to live with the consequences.

A process needs to be established as to how, and under what conditions, artificial implants can be deactivated. When a patient receives a lifesaving or life-enhancing implant, decisions about deactivation should rest with him, or if he's incapacitated, with a designated loved one, just as we do in the case of external life-support machines. Each of us should spell out our wishes in our advance directives (living wills) and doctors, nurses, hospitals, and hospices should be within the protection of a legal framework when they abide by our wishes. Most important, everyone should be educated about what to expect at the end of life when artificial organs or implants are involved. We should have prohibitions against artificially stimulating the brain when a patient is dying. We have a very long way to go before we can feel confident that we'll have a "good death," and the rapid introduction of artificial implants is a major reason why.

Although it's convenient for the purposes of discussion to make distinctions between therapies and enhancement, it's already clear that therapies will quickly morph into enhancement due to the powerful potential of converging technologies. Arguments that we should embrace therapies and reject enhancement are futile considering that there is no clear definition of normal and no bright line between a normal state and an enhanced state of being. It's more likely that once enhancement is accepted in principle, the conversation will rapidly shift to how to achieve greater and greater degrees of enhancement. After each generation of enhancements, a new normal will be established, only to be overtaken by yet another new normal. We can postulate a number of different resulting scenarios, but no one can say whether humanity will reach its biological, spiritual, and psychological limits, or if this quest will go on as long as the human race exists. We're facing monumental ambiguity in trying to predict the future of mankind. This has always been the case, of course, but as converging technologies gain momentum, the future is becoming harder and harder to predict.

Many observers predict that human nature will be fundamentally changed when radical enhancement becomes widespread. But it's also possible that converging technologies will only amplify familiar human attributes while having little to no qualitative effect. It's unlikely that converging technologies and transhumanism will eliminate many of the struggles of life; there will still be unrequited love, bereavement over the death of a loved one, competition in the economic sphere, disappointments, and just plain bad luck. But there will also be more time to recover from disappointments and to learn from our mistakes.

As we move forward into the age of transhumanism, we are also entering a new stage of technophysio evolution. Extremely long lives will necessitate a lower birthrate if we are to sustain life on this planet. Lower birthrates mean fewer heritable genetic mutations and a possible slowing of natural evolution. However, gene therapies will largely pick up where natural evolution leaves off. Humans will have greater and greater control of their biological future through gene therapies introduced into embryos.

The many technologies that are converging, rather than being studied in isolation, must be considered in their combined effect to fully understand how dramatically human life will change. After several centuries of ever-increasing specialization in the sciences, today's most forward-looking thinkers predict that only a multidisciplinary synthesis of all emerging technologies will create the renaissance of learning, thinking, and creating that lies just around the bend. Biologists must learn to think like engineers, engineers must learn to think like doctors, and doctors must learn to work with ever more powerful technologies. To create the kinds of multidisciplinary teams needed to design CT end products, our educational system must be steered away from extreme specialization and toward interdisciplinary education. Considering the intense competitive pressures of the global economy, the sooner we reform our educational system, the better.

While many decry an assault on human identity should radical enhancement become widespread, it can be argued that a definition of "human" has never been clearly laid out. Some argue that a human body with a number of artificial parts is no longer human, and neither is a chimera—a person with snippets of nonhuman DNA. With the radical and not-so-radical changes to the human body that are on the way, plus the development of androids and robots with many human features, the only way left to identify human beings as human may be shifted to the inner qualities. Things like subjective self-awareness, personality, ethical sensibility, kindness, compassion, humor, and creativity may rise in importance. We should also consider spiritual qualities like love, worship of a Creator, faith, doubt, and a certain reaction to the knowledge of our eventual mortality as fundamental to human identity. The proliferation of advanced technologies may facilitate a far more urgent examination of our spiritual natures, or in some people's views, the lack thereof.

Bioconservatives argue that if God (or nature) intended for us to have certain abilities, we would have been given (or would have evolved) them. But it can just as easily be argued that we have been endowed with the ability to help drive our own evolution. As unfinished products, we continue to evolve, and it's possible that our prolific drive to invent and utilize new technologies is an essential part of our nature, destined all along to become an integrated part of our evolution. Given its universality, our disposition of discontent and constant striving for betterment might be hardwired genetically. If this is the case, isn't it unnatural to foreclose certain advances, whether they temporarily cause social and economic disruption or eventuate in unforeseen consequences? Might the challenge of dealing with those consequences make us smarter, stronger, better than we would otherwise be?

Of real concern is the issue of distributive justice in a world of huge disparities in wealth. The widespread availability of human enhancements in wealthy nations juxtaposed with a lack of access in poor countries could be highly destabilizing and entail grave injustice. No less concerning is the wealth disparity even in rich countries. One way to ensure a just distribution of CT enhancements is through universal, nationalized medicine. This would require a major adjustment of the health care system in the United States, but the government has a compelling reason to make sure that radical life extension and advanced medical therapies are widely distributed: the health and longevity dividend. The government simply can't afford to provide Social Security and Medicare benefits to the baby boomers for several decades running, as it is now doing for their parents. Social Security was not designed to provide benefits for twenty, thirty, or more years of retirement. The age of retirement should go up with the extension of the life span, and seniors must be healthy enough to keep working long enough for the longevity dividend to offset the costs to the social safety net. Universal access to CT treatments would confer great benefits on all of society.

The prospect of radical human enhancement means that we will need to find new ways to classify illness and health, treatment and enhancement. Many people think that we already "medicalize" conditions that shouldn't be classified as a disorder—erectile dysfunction or ADHD, for example. But for regulatory and insurance purposes, our system isn't designed to deal with enhancement. The system must find a way to classify treatments that also enhance. If we don't come up with new categories for such therapies, we can expect to see the rise of many new disorders, many of them psychiatric in nature. This won't sit well with everyone. Many will argue that one shouldn't have to be labeled with a disorder in order to receive some enhancement.

To some in the bioethical community, my use of the term "bioconservative" may be seen as pejorative, but that is not my intention. The questions asked by those I designate, for lack of a better word, bioconservatives are vitally important and deserving of discussion. For full disclosure, I began this book committed to exploring all the arguments, both for and against human enhancement. In the process I have found time and again that the bioconservative arguments are less than persuasive.

It's interesting to note that, even though transhumanist thinkers trend strongly in the direction of libertarianism, many in the libertarian intelligentsia oppose their ideas. In 2009, *The New Atlantis* senior editor, Ari Schulman, argued that "the widespread use of enhancement creates tremendous social pressures to compete and conform; these pressures, too, can be said to restrict freedom."[9] While no one can argue that these pressures won't exist, as they do in any number of choices a person might make, it doesn't follow that the choice of accepting enhancement will be utterly irresistible if one wants to reject it. In an indirect way, Schulman is arguing that it's not enough to be able to reject human enhancement—one should be able to deny such choices to the rest of society lest other people's choices exert an undue influence. He seems to be saying that we should never make choices that might put pressure on others to follow suit. This argument certainly flies in the face of the libertarian leanings of *The New Atlantis,* and is not what one would expect from a libertarian thinker. He seems to be making the case that we should never make choices that another person wouldn't make, and it's our responsibility, if your choices aren't good enough for you to be happy, to make your choices happy ones.

It's probably true that those who reject enhancement once it becomes widespread will feel competitive pressures, but it's every individual's responsibility to make choices that he feels are good for him. It's not society's responsibility to limit the freedom of other people so that one can feel good about his choices. If an individual turns out to be unhappy about his choices, it's up to him to decide whether his choices are, in fact, good. Schulman argues against everyone's freedom when perhaps he should be calling for every person to have the courage to stand by his convictions. If rejecting enhancement is a good choice, it will undoubtedly come with built-in rewards.

In 2011, historian Benjamin Storey, also writing for *The New Atlantis,* said, "Should the majority decide that genetically enhancing one's children's intelligence or pharmacologically enhancing one's workplace productivity is the morally right thing to do, *not* doing so will become as taboo as smoking or failing to vaccinate one's children."[10] Storey conjures up Alexis de Tocqueville in his assertion that even in a democratic society, when one technically has the freedom to opt out of the preferences of the majority, the majority will still exercise a kind of tyranny over dissenters through social pressure. There is truth in this, but does society have the obligation to abide by the value judgments of a minority in deciding for itself what is acceptable? Both Schulman and Storey argue against society going forward with human enhancement, but fail to make a case for why those who oppose enhancement should be able to exercise free choice while those who desire it should not.

Storey writes at length to refute Ronald Bailey's assertion, in a related essay, that the arc of history is moving in a positive direction. "He does not mention," he says, ". . . that we enjoy greater longevity than our forebears at the price of an increasing number of deaths associated with the terrible realities of Alzheimer's disease: incidence of that disease is directly correlated with increases in life span, and has risen by 66 percent just in the last ten years, according to the Alzheimer's Foundation."[11] This is indeed deeply concerning, but as a historian, Storey should have acknowledged that a huge piece of this puzzle is the fact that older people now make up an unusually large segment of the population, which of course raises the incidence of AD. The *percentage* of elderly people getting AD has not changed, though people are indeed living longer. And Storey doesn't allow that the biomedical technologies he opposes may have a huge impact on the incidence of AD. He goes on to say, "the demographic crises faced by societies such as those of Italy, Spain, and Japan, could be true existential crises, and those societies could owe their fatal declines in part to the embrace of the biotechnological advancements of the twentieth century."[12] This statement is disturbing in its implication that people in these nations currently live *too* long. Would Storey really prefer a world without penicillin, open-heart surgery, and organ transplants?

Calls to abolish progress on the basis of vague feelings of repugnance and fear of the future are not likely to arrest the march of history, but that is not to say that any and all forms of human enhancement ought to be embraced without any debate. An active debate about all emerging technologies that propose to enhance human abilities should not just go on, it should widen to include more segments of society. Rather than just a narrow discussion among scientists and philosophers, the conversation should include more voices from the religious, political, environmental, legal, regulatory, and medical sectors, as well as the general public. Thinkers from the humanities should also be included to move the discussion beyond the purview of specialists. Over time, no part of society will remain untouched by the effects of human enhancement, and the more widely the issues are discussed, the better the likelihood that the social, political, and economic ecosystem will guide the process of using CTs constructively to address human needs.

Converging technologies will need appropriate oversight and regulation to ensure economic fairness and to create conditions for the widest availability of its products and services. Mihail C. Roco, who is the founding chair of the U.S. National Science and Technology Council (USNSTC) subcommittee on Nanoscale Science, Engineering and Technology, has called for both an early multidisciplinary debate on CT products and international coordination in governance to level the playing field among countries. Writing in the *Journal of Nanoparticle Research* in 2007, Roco says, "It is better to address early the long-term issues related to revolutionary implications of converging technologies in a responsible government-sponsored framework, rather than trying to adjust developments later."[13] Recognizing that the sweep of societal change due to CTs is likely to be wide, deep, and profound, he says that "the fusion of technologies is one goal. The other goal is integration of the resulting technology with human needs." He acknowledges that as technologies as powerful as nanotechnology become widespread, "some perturbations might be created that affect the very foundations of life." While we recognize that we have limited powers to predict how such perturbations might arise, we should nevertheless be actively engaged in the effort to foresee solutions to possible challenges such as social acceptance of gene therapy and sustainability in the case of overpopulation. Roco also emphasizes the need for international rules that optimize the possibility of equal access to CTs on a global level.

I spoke with Roco from his office at the National Science Foundation in Arlington, Virginia, about his thoughts on how converging technologies can be regulated on an international level. It was clear that reconciling all the cultural differences among countries to reach some coherent international policy is going to be a formidable challenge. Even the United States and most European countries meet at times with cultural differences when it comes to collaboration. But in the long run, Roco is hopeful that a convergence not just of technologies but of cultures among all the different players will come to dominate the development and implementation of CT treatments and technologies.

"It's difficult to quickly bring cultures together," he said, "but with greater interdependence, we hope, the desire to collaborate will become the dominant trend in the world." He emphasized the need for international governance as opposed to governing, explaining that it's highly likely that top-down rule making on the part of any international body is destined to fail. In fact, he said, "There is no future in the world for this." What's needed is a bottom-up movement in which individual countries voluntarily cooperate, a trend that can follow the need of scientists, engineers, and doctors throughout the world to collaborate.

I was particularly interested in what Roco thinks about the regulation of electronic health records, which, thanks to CTs, will hold a huge amount of intimate information about us. He doesn't see a panacea for cyber security, but pointed out that "opinions are evolving to accept more information sharing. Medical information needs to be shared among doctors, and this will become more and more widespread." Sharing of EHRs will enable a person to see a doctor anywhere in the world, meaning that a patient in Birmingham, Alabama, could see the world's top specialist in, say, liver cancer, through the Internet. "Information sharing could save your life," he said, "so compare that to an invasion of privacy. It's bad if someone gets your information, but it's worse if you die."

Roco noted that "in 2000 people said that Facebook would never work because it would violate privacy," yet billions of people are now perfectly happy sharing all kinds of information about themselves with the world. "Also in 2000, no one would have accepted GPS tracking them everywhere they go. Now, with cell phones, they are being tracked, and people have accepted it," he said. "The reality is if you want to be part of an interacting society, more and more of your personal information will be on computers." This is hardly reassuring, but it underlines the extreme difficulty, and perhaps impossibility, of ever truly protecting the vast amount of electronic information about us that can be hacked. It also illustrates how, when choosing between privacy and convenience, we seem to be much more inclined to go for convenience, and this could be our Achilles' heel when it comes to being cautious about new technologies.

While others seek to sort out the regulatory issues, the ethical debate about human enhancement is no less urgent. The Swedish philosopher and transhumanist Anders Sandberg outlined in 2001 his call for what he terms "morphological freedom," or the freedom to change or not to change our bodies as we choose. Sandberg says, "That individuals have rights does not absolve them from their obligations to each other or their need of each other. But these obligations and needs cannot ethically override the basic rights. No matter what the social circumstances are, it is never acceptable to overrule someone's right to life or morphological freedom. For morphological freedom—or any other form of freedom—to work as a right in society we need a large dose of tolerance."[14]

Tolerance is indeed the social quality needed to protect both those who find self-expression in human enhancement and those who reject it. This principle has been essential since the first establishment of democratic forms of government, and happily, democratic societies are moving in the direction of accepting more diversity. Women and minorities of all stripes are gaining more entrance and acceptance into all sectors of society. Enhancing one's self in myriad ways has always been a form of self-expression, and as more technologies become cheaper and more widely available, they may take self-expression to heights never seen before. It's likely that a future society will include many shades of human-artificial hybrids. It may be that the concept of personhood, after being bestowed on a greater variety of beings, will be distinguished from humanity. The definition of "human" is likely to undergo revision if not a complete redefinition, and at some point in the far future, we may cease to call ourselves part of the species Homo sapiens. While many people today are disturbed by these notions, once we have traveled far enough on the transhuman journey, we may have a completely different assessment of them. And, we could end up clearly seeing what traits in human beings are the most enduring and fundamental—at last, a definition of the true human essence.

The Australian ethicist Nicholas Agar, in his 2011 book *Humanity's End: Why We Should Reject Radical Enhancement*, frequently turns to nature as his touchstone for determining what is natural and therefore desirable. Nature does indeed warrant reverence and consideration, but even in nature, no species lasts forever. If Neanderthals had had the power to perpetuate their species forever, they probably would have done so, but modern humans can hardly regret the fact that they became extinct and made way for us. Considering that human evolution is still ongoing, we can't say whether Homo sapiens is the final endpoint for humanity. But given that we are taking greater control of our evolution through emerging technologies, we may have a deciding role to play in what we become.

Sandberg observes, "But even if one accepts the idea of a particular human nature this nature seems to include self-definition and a will to change as important aspects; a humanity without these traits would be unlike any human culture ever encountered. It is rather *denying* these traits to oneself or others that would go against human nature."[15] By ignoring the human will to change, which looms so large in our individual and collective lives, bioconservatives fail to make compelling arguments against transhumanism. According to Sandberg, the search for continuous self-transformation is an enduring trait, one that will inevitably change our perceptions as we attain new levels of existence. Hence, while humans today might elect to make all their children tall, blond, and athletic, those children might value entirely different traits in their children. It's reasonable to consider that when people have more control over their physical bodies, traits other than appearance might rise in importance. Sandberg proposes that society establish a strong commitment to morphological freedom. This would protect those who wish to reject enhancement just as much as it protects those who want to embrace it; it would make it difficult to enforce some kind of compulsory behavior either for or against enhancement.

One issue that seems to receive little attention in discussions about morphological freedom is the impact on relationships should we decide to radically change our bodies. Fundamental relationships such as marriage could require major adjustments, considering that one enters such relationships while a person is in possession of a certain body, and we tend to get deeply attached to our loved ones' bodies. Society would have to go through adjustments to accept morphological freedom, but in a society where morphological freedom is embraced, we would necessarily know our loved ones for their inner qualities, knowing that their bodies could change dramatically over the course of a lifetime. On the other hand, there will likely be limits to human enhancement. The point is to create healthier, smarter, more beautiful and long-lived humans, not to change a human being into a jellyfish.

As welcome as the radical cures for disease are to most of us, we can't assume that everyone will want to embrace them. Sandberg points to members of the deaf community who reject cochlear implants as an example of those whose identities are so intertwined with a disability that they would regard a cure as a loss of dignity, identity, and community. "As the lines blur between curative and augmentative treatments," he says, "self-expression moves further into the realm of self-transformation and treatments that might be desirable by some people but not to others (such as cochlear hearing implants or genetic therapy) become available, it becomes increasingly hard to define what constitutes a natural body and what is a body modified in a volitional way. . . . Taking the step to full morphological freedom creates a far simpler ethical guideline, which both protects those who do not wish to change, those who are differently bodied and those wanting to change their bodies."[16] As for how morphological freedom would affect our definition of humanity, he says, "From my perspective morphological freedom is not going to eliminate humanity—but to express what is truly human even further."[17]

The concept of morphological freedom fits comfortably within the core values of a democratic society, but no system can eradicate all discontent. While we can and should avoid personal choices that inflict direct harm on others, society can't be responsible for the harm that some people will choose to inflict on themselves. The best that society can do is to protect everyone's right to act freely in their own self-interest. While we might hope for a world in which no one suffers the consequences of harmful choices, there are limits to what a democratic and free society can do.

Nicholas Agar, in *Humanity's End*, argues that rather than fully embracing or banning human enhancement, there is a third way—accepting only moderate enhancement and banning radical enhancement. Our definition of what constitutes moderate enhancement would be the highest level achieved naturally by a member of our species. For example, the longest human life ever recorded is 122 years, so presumably we should limit ourselves to a 122-year life span. But Agar's proposal quickly meets with trouble. What do we do when the first person turns 123? Also, who would be the arbiter of what constitutes radical enhancement when technologies continue to evolve, and how would a ban on radical enhancement be enforced?

The line between treatment and enhancement is already blurring, and as technology advances, our view of what's normal will continue to change. There is currently no consensus about what constitutes normality in almost any human trait. Furthermore, the core values of all democratic societies would be violated if we were to ban some enhancement that some people consider radical and others do not. All bans would have to rest on arbitrary determinations such as a maximum life span or a maximum level of intellectual ability (measured how?), but these arbitrary judgments would quickly become obsolete as technology continues to advance. For the government or any other entity to enforce such bans, the rights to privacy and autonomy would have to be violated on a scale never before allowed in any democratic society. The necessary violation of these rights would be a precedent for possibly many other abuses—in fact, the practice of enforcing a ban on radical enhancement would effectively destroy a whole spate of constitutionally guaranteed rights, including the basic right to life, liberty, and the pursuit of happiness. In essence, we could not ban the right to embrace radical enhancement and maintain a democracy at the same time.

Agar and many bioconservatives also overlook a glaring problem when it comes to enforcing the bans they call for, at least in the United States. Because medicine is largely delivered through the for-profit private sector, patients are also customers, and health care deliverers are businesses. Questions that pertain to the maintenance of free markets would come into play. Even if radical enhancement were in the beginning available only to those who can afford it, many people would want it, and deliverers would want to sell it. The profit-seeking nature of biotech companies, pharmaceutical companies, medical device manufacturers, hospitals, doctors, and technicians are enough of an incentive for them to push continually for the freedom to make technological advances available to the largest number of people. These constituencies have a great deal of lobbying power in Washington, and they stand to make unheard-of profits on enhancing technologies. They will continue to push for the right to benefit from NIH-funded research, to patent their inventions, to conduct business in the United States, and to sell new technologies and treatments. Banning what some deem to be radical enhancements would turn the American health care system, and democracy itself, on its head.

Countries that ban radical enhancements will find themselves at a competitive disadvantage to those that support them. Those nations that make such treatments available will potentially see a great increase in medical tourism, as those who can afford enhancement but who live in countries that ban them would still have access.

If we were to attempt to ban human enhancement, even more fundamental would be the radical shift in our philosophical orientation toward human progress. As noted above, human beings have sought continuous self-improvement since the beginning of recorded history. Given the persistence and universality of this phenomenon, it can be argued that the longing for self-improvement is genetically hardwired into our species. It certainly pervades every culture, and for us to suddenly take a different stance does not seem realistic. It is for the following reasons that the widespread adoption of radical enhancement may already be inevitable.

People have started to embrace medical implants and a large array of advanced medical technologies, and this phenomenon has established significant momentum. The lack of an agreed-upon definition of normal makes it impossible to draw lines between health and disease and between therapy and enhancement. Through life-extending technologies and treatments, we have already blurred the boundaries between medical therapies and enhancements, and this line will only get blurrier with time. Industrialized democracies tend to have entrenched cultures that accept and protect individual choice, and the banning of radical enhancement would turn democratic values on their head. It would take a drastic reorientation to convince citizens of a democracy who are also customers in a free-market economy that they can't receive treatments they desire and that were developed, to a great extent, with their tax dollars.

As our bodies become ever more artificial, we will increasingly make a distinction between human biology and human nature. Those who think that enhancing our bodies and brains will change human nature, whether they realize it or not, are grounding human nature in biology. But it remains to be seen whether augmenting our bodies does, in fact, change human nature. The technologies that exist today, or are under development, amplify abilities that humans already have—this we know for sure. Whether there is a tipping point in human intellectual ability or any other ability when we will become something qualitatively different from human remains to be seen.

Many bioconservatives couch their objections to human enhancement in terms that suggest that technology is in competition with religion. There's a vague assumption that when technology brings human enhancement and even possibly immortality, humans will no longer need God. But this assumption also rests on the unarticulated assumption that technology can somehow satisfy all the deep longings of humanity for meaning and transcendence, an assumption that not all transhumanists entertain. Having explored the writings of several leading bioconservatives as I researched this book, it has become more and more apparent that it is the conservative ethicists who seem to lack conviction that God is truly omnipotent or that human nature is more than the human biological organism. The idea that technology has the power to transform the spiritual condition, to fill all of our longings for happiness and perfection, has not been proven and needs far more discussion before we assume that it does. But that doesn't mean it couldn't make life richer, deeper, and more meaningful than ever.

It's time to advance the discussion of human enhancement beyond the impasse of the past few decades. Science has overtaken futile arguments about playing God with an explosion of CT innovation. We need to get used to the idea that we are co-creators of ourselves. If we fail to accept the inevitability of transhumanism and decline to take conscious control of it, we're abdicating our responsibilities on a fundamental level. In addition, the serious social and ethical issues that must be sorted out won't go away as we continue to have futile arguments about whether or not mankind should control its destiny or leave everything important up to chance. Mankind long ago started controlling its destiny, and the only difference is that now we have truly powerful tools to aid in that enterprise. Given the rapidly advancing state of technology, the conversation should not be about whether we have the courage to control our destiny, it should be about how we do so.

One can look at the human enhancement project as a gigantic human experiment, as bioconservatives already do. They caution that we can't know the outcome of this experiment in advance, but converging technologies will allow us to examine human nature as never before. In the last analysis, any attempt to predict the consequences of human enhancement will fall short. We simply have too little experience to make accurate predictions, but we can gain clues by reflecting on how society has been changed by technophysio evolution, and by how human beings fare psychologically as they reach older ages. While these trends haven't created a paradise on earth, in most ways life has gotten better—longer, healthier, and more prosperous, at least. People have far more time to reflect, to enjoy loved ones, and to pursue self-actualization. We might cautiously extrapolate that this general trend will continue.

There are a huge number of questions that we may never be able to answer until we journey further into human enhancement. No one has yet arrived at an accepted definition of what it means to be human. Like each of us, that definition is in flux, and will likely remain in flux for the foreseeable future. In the end, what we want to become may define us more than where we've been. As philosophers and scientists continue to debate what constitutes human nature, we may only ever see who we are today in the rearview mirror, from a state far more advanced than where we find ourselves now.

# Notes

## 1. WHEN HUMANS AND TECHNOLOGY MERGE

1. Bailey, Ronald, "Transhumanism: The Most Dangerous Idea? Why Striving to Be More Than Human Is Human," *Reason*, accessed August 5, 2013, http://reason.com/archives/2004/08/25/transhumanism-the-most-dangero.

2. Francis Fukuyama, "Transhumanism," *Foreign Policy*, September 1, 2009, accessed March 2, 2016, http://foreignpolicy.com/2009/10/23/transhumanism/.

3. Anders Sandberg, "Morphological Freedom—Why We Not Just Want it, but *Need* It," accessed August 23, 2013, http://www.aleph.se/Nada/Texts/MorphologicalFreedom.htm.

4. Courtney S. Campbell et al., "The Bodily Incorporation of Mechanical Devices: Ethical and Religious Issues," part I, *Cambridge Quarterly of Healthcare Ethics* 16 (2007): 227–37.

5. Mark S. Frankel and Cristina J. Kapustij, "Enhancing Humans," The Hastings Center, accessed September 5, 2013, http://thehastingscenter.org/Publications/BriefingBook/Detail.aspx?id=2162&terms=n.

6. Y. J. Erden, "ICT Implants, Nanotechnology and Some Reasons for Caution," BioCentre, accessed March 2, 2016, http://www.bioethics.ac.uk/news/ICT-Implants-nanotechnology-and-some-reasons-for-caution.php.

7. Nick Bostrom, "A History of Transhumanist Thought," *Journal of Evolution and Technology* 14, no. 1 (2005): 7.

8. Ibid., 12.

## 2. "BETTER THAN THE HEART I WAS BORN WITH"

1. The National Network of Organ Donors, accessed July 1, 2013, http://www.organdonor.gov/index.html.

2. L. A. Jansen, "Hastening Death and the Boundaries of the Self," *Bioethics* 20 (2006): 105–11, doi:10.1111/j.1467-8519.2006.00481.x.

3. Rob Stein, "Devices Can Interfere with Peaceful Death," *The Washington Post,* December 17, 2006.

4. Erden, "ICT Implants, Nanotechnology, and Some Reasons for Caution."

5. James E. Russo, "Deactivation of ICDs at the End of Life: A Systematic Review of Clinical Practices and Provider and Patient Attitudes," *American Journal of Nursing* 3, no. 10 (October 2011): 32.

6. Nathan E. Goldstein et al., "That's Like an Act of Suicide: Patients' Attitudes Toward Deactivation of Implantable Defibrillators," *Journal of General Internal Medicine* 23, supplement 1 (January 2008): 7–12, accessed on July 22, 2013, http://www.ncbi.nlm.nih.gov/pmc/articles/PMC2150628/.

7. Nathan E. Goldstein et al., "It's Like Crossing a Bridge: Complexities Preventing Physicians from Discussing Deactivation of Implantable Defibrillators at the End of Life," *Journal of General Internal Medicine* 23, supplement 1 (January 2008): 2–6, accessed July 22, 2013, http://www.ncbi.nlm.nih.gov/pmc/articles/PMC2150631/.

8. Russo, "Deactivation of ICDs at the End of Life," 29.

9. Richard A. Zellner et al., "Controversies in Arrhythmia and Electrophysiology: Should Implantable Cardioverter-Defibrillators and Permanent Pacemakers in Patients with Terminal Illness Be Deactivated?" *Circulation: Arrhythmia and Electrophysiology* 2 (2009): 340–44.

10. Daniel P. Sulmasy, "Within You/Without You: Biotechnology, Ontology, and Ethics," *Journal of General Internal Medicine* 23, supplement 1 (January 2008): 69–72.

11. Ibid.


## 3. THE RACE TO BEAT KIDNEY, LUNG, AND LIVER DISEASE

1. D. Martins, N. Tareen, and K. C. Norris, "The Epidemiology of End-Stage Renal Disease Among African Americans," *American Journal of Medical Science* 323, no. 2 (February 2002): 65–71.

2. Organ Procurement and Transplantation Network, accessed March 2, 2016, https://optn.transplant.hrsa.gov/.

3. Erin Allday, "Kidney Designers Take Cues from Nature," SFGate, accessed March 2, 2016, http://www.sfgate.com/health/article/Kidney-designers-take-cues-from-nature-4458059.php.

4. Yosuke Shimazono, "The State of the International Organ Trade: A Provisional Picture Based on Integration of Available Information," *Bulletin of the World Health Organization,* accessed October 23, 2013, http://www.who.int/bulletin/volumes/85/12/06-039370/en/.

5. Larry Rohter, "The Organ Trade: A Global Black Market; Tracking the Sale of a Kidney on a Path of Poverty and Hope," *The New York Times,* May 23, 2004, accessed March 2, 2016, http://www.nytimes.com/2004/05/23/world/organ-trade-global-black-market-tracking-sale-kidney-path-poverty-hope.html?_r=0.

6. Eli A. Friedman, "Ethical Stresses in Uremia Therapy: The Worst Is Yet to Come," *ASAIO Journal* 48, no. 3 (May/June 2002): 209–10.

7. Delicia Honan Yard, "Implantable Artificial Kidney Could Help Tens of Thousands: Interview with Shuvo Roy, Ph.D.," *Renal & Urology News,* accessed March 2, 2016, http://www.renalandurologynews.com/expert-qu/implantable-artificial-kidney-could-help-tens-of-thousands-interview-with-shuvo-roy-phd/article/272364/.

8. University of California, San Francisco, Schools of Pharmacy and Medicine, accessed July 15, 2014, http://pharmacy.ucsf.edu/kidney -project/device.

9. Yard, "Implantable Artificial Kidney Could Help Tens of Thousands," 4.

10. Shuvo Roy and William Fissell, "The Kidney Project FAQ, Version 2.0," accessed March 2, 2016, http://pharm.ucsf.edu/sites/pharm .ucsf.edu/files/kidney/media-browser/Patient%20FAQ%20-%20 July%203%2C%202014.pdf.

11. "The Debate About Converging Technologies," *Nanotechnology Spotlight,* accessed November 15, 2013, http://www.nanowerk.com /spotlight/spotid=6569.php.

12. Mihail C. Roco and William Sims Bainbridge, eds., "Converging Technologies for Improving Human Performance: Nanotechnology, Biotechnology, Information Technology and Information Science," accessed August 19, 2013, http://www.wtec.org/Converging Technologies/Report/NBIC-report.pdf.

13. "The Debate About Converging Technologies."

14. Ibid.

15. Achilles A. Demetriou et al., "A Bioartificial Liver to Treat Severe Acute Liver Failure," *Annals of Surgery* 239 (2004): 660–70.

16. Sherwin B. Nuland, *How We Die: Reflections on Life's Final Chapter* (New York: Alfred A. Knopf, 1993), 67.

17. Nathan Longtin, "MC3's BioLung," accessed October 19, 2013, http://www.ele.uri.edu/courses/bme181/F08/Nate_1.pdf.

18. Francis Fukuyama, *Our Posthuman Future: Consequences of the Biotechnology Revolution* (New York: Farrar, Straus and Giroux, 2002), 9.

19. Ibid., 101.

## 5. JUMP-STARTED BY THE U.S. MILITARY

1. Gali Halevi, "Military Medicine and Its Impact on Civilian Life," *Research Trends,* accessed March 17, 2015, http://www.research trends.com/issue-34-september-2013/military-medicine-and-its -impact-on-civilian-life/.

2. Allison Lex, "Nine Things Invented for Military Use That You Now Encounter in Everyday Life," *Mental Floss,* accessed March 7, 2015, http://mentalfloss.com/articles/31510/9-things-invented-military -use-you-now-encounter-everyday-life.

3. Kenneth Chang, "Scotch Tape Unleashes X-Ray Power," *The New York Times,* October 23, 2008.

4. Roco and Bainbridge, eds., "Converging Technologies for Improving Human Performance."

5. Antonio Regalado, "Military Funds Brain-Computer Interfaces to Control Feelings," *MIT Technology Review,* accessed March 7, 2015, http://www.technologyreview.com/news/527561/military-funds -brain-computer-interfaces-to-control-feelings.

6. James Hughes, *Citizen Cyborg: Why Democratic Societies Must Respond to the Redesigned Human of the Future* (Cambridge, MA: Westview Press, 2004), 128.

7. Ibid., 129.

8. Ronald Bailey, "The Case for Enhancing People," *The New Atlantis,* summer 2011, accessed March 13, 2015, http://www.thenewatlantis .com/publications/the-case-for-enhancing-people.

9. American Psychiatric Association, "Paraphilic Disorders," in *Diagnostic and Statistical Manual of Mental Disorders, Fifth Edition* (Arlington, VA: APA, 2012), 685–705.

10. Ethan A. Huff, "The United States of Plastic Surgery: Americans Spent $11 Billion Last Year on Face Lifts, Botox, Breast Augmentations," Natural News, accessed March 2, 2016, http://www.natural news.com/040164_plastic_surgery_breast_augmentation_Botox .html.

11. Bruno Macaes, "Technology and Authenticity," *The New Atlantis,* accessed March 13, 2015, http://www.thenewatlantis.com/publications /technology-and-authenticity.

12. Ibid.

13. Ibid.

14. Ibid.

15. Bailey, "The Case for Enhancing People."

## 6. BUILDING A BETTER BRAIN

1. Pam Belluck, "Dementia Care Cost Is Projected to Double by 2040," *The New York Times,* accessed April 4, 2013, http://www.nytimes.com/2013/04/04/health/dementia-care-costs-are-soaring-study-finds.html.

2. Constantine G. Lyketsos et al., "Deep Brain Stimulation: A Novel Strategy for Treating Alzheimer's Disease," *Innovation in Clinical Neuroscience* 9, no. 11–12 (November–December 2012): 13.

3. Ibid., 12.

4. Todd Essig, "Scientists Call Foul on Brain Games Pseudo-Science," *Forbes*, accessed February 29, 2016, http://www.forbes.com/sites/toddessig/2014/10/29/scientists-call-foul-on-brain-games-pseudo-science/#be9e3665e1ad.

5. James Hughes, "Enhancing Virtues: Intelligence (Part 4): Brain Machines," Institute for Ethics and Emerging Technologies, September 22, 2014, accessed March 19, 2015, http://ieet.org/index.php/IEET/more/hughes20140922.

6. Ronald Bailey, "The Battle for Your Brain," Center for Cognitive Liberty & Ethics, 2003, accessed March 19, 2015, http://www.cognitiveliberty.org/neuro/bailey.html.

7. Ibid.

8. Ibid.

9. Ibid.

10. Ibid.

11. Ibid.

12. Margaret Talbot, "Brain Gain: The Underground world of 'Neuroenhancing' Drugs," *The New Yorker,* April 27, 2009, accessed April 4, 2015, http://www.newyorker.com/magazine/2009/04/27/brain-gain.

13. A. K. Brem et al., "Is Neuroenhancement by Noninvasive Brain Stimulation a Net Zero-Sum Proposition?" *Neuroimage* 85: 1058–68, doi:10.1016/jneuroimage.2013.07.038.

14. Ibid.

15. Helen S. Mayberg et al., "Deep Brain Stimulation for Treatment-Resistant Depression," *Neuron* 45 (March 3, 2005): 651–60.

16. Hughes, "Enhancing Virtues," 3.

17. Kenneth Hayworth, "Killed by Bad Philosophy: Why Brain Preservation Followed by Mind Uploading Is a Cure for Death," The Brain Preservation Foundation, January 2010, accessed March 18, 2015, http://brainpreservation.org/content/killed-bad-philosophy.

18. Michelle Croteau, "Less Than One in Three Americans Have a Living Will, Says New FindLaw.com Survey," *PR Newswire,* http://prnewswire.com/news-releases/less-than-one-in-three -americans-have-a-living-will-says-new-findlawcom-survey -190163891.html.

## 7. THE AGELESS SOCIETY

1. American Society of Plastic Surgeons, "2014 Plastic Surgery Report," accessed March 1, 2016, http://www.plasticsurgery.org/Documents /news-resources/statistics/2014-statistics/plastic-surgery-statsitics -full-report.pdf.

2. Ibid.

3. Elizabeth O'Brien, "Ten Secrets of the Anti-Aging Industry," MarketWatch, 2014, accessed April 23, 2015, http://www.marketwatch .com/story/10-things-the-anti-aging-industry-wont-tell-you-2014 -02-11.

4. Gregg Easterbrook, "What Happens When We All Live to 100?" *The Atlantic,* October 2014, accessed April 23, 2015, http://www .theatlantic.com/features/archive/2014/09/what-happens-when-we -all-live-to-100/379338/.

5. Ian Sample, "Harvard Scientists Reverse the Ageing Process in Mice," *The Guardian,* November 28, 2010, accessed April 23, 2015, http://www.theguardian.com/science/2010/nov/28/scientists -reverse-ageing-mice-humans.

6. Easterbrook, "What Happens When We All Live to 100?"

7. INSERM, "Erasing Signs of Aging in Human Cells Now a Reality," *Science Daily,* November 7, 2011, accessed July 24, 2012, http://www .sciencedaily.com/releases/2011/11/111103120605.htm.

8. NIH/National Heart, Lung and Blood Institute, "Single Gene Change Increases Mouse Lifespan by 20 Percent," *Science Daily,* August 29,

2013, accessed August 30, 2013, http://www.sciencedaily.com
/releases/2013/08/130829124011.htm?utm_source=feedburner+utm
_medium=email+utm_campaign=Feed%3A+sciencedaily%Ftop
-science+%28ScienceDaily%3A+Top+News+—+Top+Science%29.

9. Ed Regis and George Church, "The Recipe for Immortality: An Expert in Synthetic Biology Explains How People Could Soon Live for Centuries," *Discover Magazine,* October 17, 2012, accessed April 20, 2015, http://discovermagazine.com/2012/oct/20-the-recipe-for -immortality.

10. Bill Gifford, "Does a Real Anti-Aging Pill Already Exist? Inside Novartis's Push to Produce the First Legitimate Anti-Aging Drug," *Bloomberg Business,* February 12, 2015, accessed March 2, 2016, http://www.bloomberg.com/news/features/2015-02-12/does-a-real -anti-aging-pill-already-exist-.

11. Ibid.

12. Bradley J. Fikes, "Anti-Aging Drugs Discovered," *The San Diego Union-Tribune,* March 9, 2015, accessed April 23, 2015, http://www .utsandiego.com/news/2015/mar/09/aging-scripps-mayo -senolytics/.

13. Gian Volpicelli, "Meet Aubrey de Grey, the Researcher Who Wants to Cure Old Age," *Motherboard,* May 23, 2014, accessed April 20, 2015, http://motherboard.vice.com/read/meet-aubrey-de-grey-the -researcher-who-wants-to-cure-old-age.

14. Ker Than, "Hang in There: The 25-Year Wait for Immortality," *Live Science,* April 11, 2005, accessed April 20, 2015, http://livescience .com/6967-hang-25-year-wait-immortality.html.

15. Kathleen Caulderwood, "Big Pharma Makes Headway on 'Fountain of Youth' Drug, Pouring Money, Resources Into Anti-Aging," *International Business Times,* December 30, 2014.

16. K. Eric Drexler, "Engines of Creation: The Coming Era of Nanotechnology," chapter 7, accessed April 20, 2015, http://e-drexler.com/d /06/00/EOC_Chapter_7.html.

17. Ibid.

18. Pew Research Center, Religion and Public Life, "To Count Our

Days: The Scientific and Ethical Dimensions of Radical Life Extension," August 6, 2013, accessed December 20, 2013, http://www
.pewforum.org/2013/08/06/to-count-our-days-the-scientific-and
-ethical-dimensions.

19. Robert W. Fogel and Dora L. Costa, "A Theory of Technophysio Evolution, with Some Implications for Forecasting Population, Health Care Costs and Pension Costs," *Demography* 34, no. 1 (February 1997): 49–66.

20. Ibid.

21. Ibid.

22. Pew Research Center, Religion and Public Life, "Living to 120 and Beyond: Americans' Views on Aging, Medical Advances and Radical Life Extension," August 6, 2013, accessed October 8, 2013, http://www.pewforum.org/2013/08/06/living-to-120-and-beyond
-americans-views-on-aging.

23. Harvard University, "Living Longer, Living Healthier: People Are Remaining Healthier Later in Life," *Science Daily,* accessed July 30, 2013, http://www.sciencedaily.com/releases/2013/07/130729083352
.htm?utm_source=feedburner&utm_medium=email&utm_
campaign-Feed%3A+sciencedaily%3A+Top+News+—+Top
+Science%29.

24. Laura L. Carstensen, "Growing Old or Living Long: Take Your Pick," *Issues in Science and Technology,* accessed April 23, 2015, http://issues
.org/23-2/carstensen/.

25. Ibid.

26. Ibid.

27. F. Schmiedek, M. Lovden, and U. Lindenberger, "Keeping It Steady: Older Adults Perform More Consistently on Cognitive Tests than Younger Adults," *Psychological Science,* 2013, doi:10.1177/09567976613479611.

28. Association for Psychological Science, "Young Versus Old: Who Performs More Consistently?" *Science Daily,* August 5, 2013, accessed August 6, 2013, http://sciencedaily.com/releases/2013/08
/130805223438.htm?_source=feedburner&utm_medium=email

+utm_campaign=Feed%3A+sciencedaily%2Ftop_health+%28Sci
enceDaily%3A+top+News+—+Top+Health%29.

29. Michael Ramscar et al., "The Myth of Cognitive Decline: Non-linear Dynamics of Lifelong Learning," *Topics in Cognitive Science,* 6 (January 2014): 5–42.

30. Easterbrook, "What Happens When We All Live to 100?"

31. S. Jay Olshansky et al., "In Pursuit of the Aging Dividend," *The Scientist,* March 2006, 31.

32. Easterbrook, "What Happens When We All Live to 100?"

33. John K. Davis, "Life Extension and the Malthusian Objection," *Journal of Medicine and Philosophy* 30 (2005): 27–44.

34. Cato Unbound, "Do We Need Death? The Consequences of Radical Life Extension," December 3, 2007, accessed April 24, 2015, http://www.cato-unbound.org/2007/12/03/aubrey-de-grey/old-people-are-people-too-why-it-our-duty-fight-aging-death.

35. Diana Schaub, "Ageless Mortals," Cato Unbound, accessed April 20, 2015, http://www.cato-unbound.org/2007/12/05/diana-schaub/ageless-mortals.

36. Ibid.

37. Ronald Bailey, "Do We Need Death?" Cato Unbound, accessed April 24, 2015, http://www.cato-unbound.org/2007/12/07/ronald bailey/do-we-need-death.

38. Daniel Callahan, "Nature Knew What It Was Doing," Cato Unbound, accessed April 24, 2015, http://cato-unbound.org/2007/12/10/daniel-callahan/nature-knew-what-it-was-doing,

39. Bailey, "Do We Need Death?"

40. Callahan, "Nature Knew What It Was Doing."

## 8. THE AGE OF SOCIAL ROBOTS

1. "Healthcare Robotics 2015–2020: Trends, Opportunities and Challenges," *Robotics Business Review,* accessed June 25, 2015, http://www.roboticsbusinessreview.com/research/report/healthcare_robotics_2015_2020_trends_opportunities_challenges.

2. Jonathan Cohn, "The Robot Will See You Now," *The Atlantic,*

March 2013, accessed January 27, 2014, http://theatlantic.com/magaz
ine/print/2013/03/the-robot-will-see-you-now/309216/.

3. Ibid.

4. Bill Gates, "A Robot in Every Home," *Scientific American,* January
2007, accessed June 25, 2015, http://www.scientificamerican.com
/article/a-robot-in-every-home/.

5. Alexis C. Madrigal, "Meet the Robotics Company Apple Just An-
nointed," *The Atlantic,* June 13, 2013, accessed July 2, 2015, http://
www.theatlantic.com/technology/archive/2013/06/meet-the
-robotic-company-apple-just-annointed/276860/.

6. Ibid.

7. Ibid.

8. Ibid.

9. Doug Cameron and Alistair Barr, "Google Snubs Robotics Rivals,
Pentagon," *The Wall Street Journal,* March 5, 2015, accessed July 1,
2015, http:///www.wsj.com/articles/google-snubs-robotics-rivals-pen
tagon-1425580734.

10. Celeste LeCompte, "Inside Google's Latest Series of Acquisitions,"
*Robotics Business Review*, December 6, 2013, accessed March 2,
2016, http://www.roboticsbusinessreview.com/article/inside_googles
_latest_series_of_acquisitions.

11. Georgia Institute of Technology, "Putting a Face on a Robot," *Science
Daily,* October 1, 2013, accessed October 2, 2013, http://sciencedaily
.com/releases/2013/10/131001104543.htm?utm_source=feedburner
&utm_medium=email&utm_campaign=Feed%3A+Top+News+—
+Top+Technology%29.

12. Jenay M. Beer et al., "The Domesticated Robot: Design Guidelines
for Assisting Older Adults to Age in Place," Institute of Electrical and
Electronics Engineers, Proceedings of the Seventh Annual ACM/
IEEE Conference on Human-Robot Interaction, March 5–8, 2012,
Boston, MA.

13. "Japan—Robots As Caregivers for Frail Parents?" Suite 101.com,
September 29, 2009, accessed August 22, 2012, http://suite101com
/article/service-robots-as-caregivers-for-frail-elderly-a153372.

14. Ibid.

15. European Commission, CORDIS, "A Personalized Robot Companion for Older People," *Science Daily*, August 16, 2013, accessed August 19, 2013, http://sciencedaily.com/releases/2013/08/1308161 25631.htm?

16. Ibid.

17. Hari Kunzru, "You Are Cyborg," *Wired*, February 1, 1997, accessed July 22, 2013, http://www.wired.com/wired/archive/5.02 /ffharaway_pr.html.

18. John Frank Weaver, "Robots Are People, Too," *Slate*, July 27, 2014, accessed June 17, 2015, http://www.slate.com/articles/technology /future_tense/2014/07/ai_drones_ethics_and_laws_if_corpora tions_are_people_so_are_robots.html.

19. Mark Goldfeder, "The Age of Robots Is Here," 2045 Strategic Social Initiative, March 2, 2016, http://2045.com/news/32949.html.

20. Patrick Tucker, "The Military Wants to Teach Robots Right from Wrong," *The Atlantic*, May 14, 2014, accessed June 17, 2015, http:// www.theatlantic.com/technology/archive/2014/05/the-military -wants-to-teach-robots-right-from-wrong/370855.

## 9. JUST DON'T CALL IT TRANSHUMANISM

1. Francis Fukuyama, "Transhumanism," *Foreign Policy*, October 23, 2009, accessed April 11, 2016, http://foreignpolicy.com/2009/10/23 /transhumanism/.

2. Ibid.

3. Leon Kass, "The Wisdom of Repugnance," *The New Republic*, June 2, 1997, 22.

4. Pew Research Center, Religion and Public Life, "Religious Leaders' Views on Radical Life Extension," August 6, 2013, accessed April 20, 2015, http://www.pewforum.org/2013/08/06/religious-leaders-views -on-radical-life-extension/.

5. Ibid.

6. Ibid.

7. Ibid.

8. Ibid.

9. Ari N. Schulman, "The Myth of Libertarian Enhancement," *The New Atlantis,* October 14, 2009, accessed March 13, 2015, http://futurisms.thenewatlantis.com/2009/10/myth-of-libertarian-enhancement.html.

10. Benjamin Storey, "Libertarian Biology, Lost in the Cosmos," *The New Atlantis,* summer 2011, accessed May 23, 2015, http://www.thenewatlantis.com/publications/liberation-biology-lost-in-the-cosmos.

11. Ibid.

12. Ibid.

13. Mihail C. Roco, "Possibilities for Global Governance of Converging Technologies," *Journal of Nanoparticle Research,* doi:10.007/s11051-007-9269-8.

14. Sandberg, "Morphological Freedom."

15. Ibid.

16. Ibid.

17. Ibid.

# Index